NUCLEAR POWER,
MAN AND THE ENVIRONMENT

The Wykeham Science Series

General Editors:

PROFESSOR SIR NEVILL MOTT, F.R.S.

Emeritus Cavendish Professor of Physics
University of Cambridge

G.R. NOAKES

Formerly Senior Physics Master
Uppingham School

The Author:

R. J. PENTREATH graduated from Queen Mary College, University of
London, before moving to the University of Auckland, New Zealand,
to carry out zoological research for his Ph.D. Since 1969 he has been
employed as a radiobiologist at the MAFF Fisheries Radiobiological
Laboratory, Lowestoft, and has published many papers on radionuclide
accumulation by aquatic organisms as well as on pathways of radionuclide
transfer to man via food chains, for which he was awarded a D.Sc. by the
University of London.

The Teacher:

M. I. SMITH has taught biology and general science at all secondary school
levels, and is now part-time sixth-form mistress at Norwich High School
for Girls, where she herself was educated.

NUCLEAR POWER, MAN AND THE ENVIRONMENT

R.J. PENTREATH

Fisheries Radiobiological Laboratory, Lowestoft

TAYLOR & FRANCIS LTD, LONDON
1980

First published 1980 in the Wykeham Science Series by Taylor & Francis Ltd.
4 John St. London, WC1N 2ET

British Library Cataloguing in Publication Data

Pentreath, R J
  Nuclear power, man and the environment. –
  (Wykeham science series; 51).
  1. Radioactivity – Physiological effect
  2. Atomic power-plants – Environmental aspects
  I. Title
  614.7   QP82.2.153   80-20173

  ISBN 0-85109-840-1

Phototypesetting by Georgia Origination, Liverpool.
Printed in Great Britain by Taylor & Francis (Printers) Ltd.,
Rankine Road, Basingstoke, Hants RG24 0PR

Preface

Nuclear power has had a chequered history, both in its development and in its acceptance by the public. For some it represents the best possible choice of energy production in the future; for others it is something to be avoided at all costs. Why such extremes of views? Part of the reason must lie in the means by which the world first became aware of nuclear energy – the explosion of the atomic bombs on Hiroshima and Nagasaki. The fear that a nuclear reactor is a potential atomic bomb is still a very common misconception. More recently, concern has centred not so much on the nuclear reactor itself as on its fuel cycle. Whereas a nuclear reactor does not constitute a nuclear weapon, its spent fuel contains plutonium which, in the right form, can be used to create one. Plutonium is also rather a hazardous substance – one of many – which results from the use of nuclear power.

There are books which argue for, and books which argue against, nuclear power; there are books which compare different potential sources of energy, and books which even question the continuing need for more energy in our society. This book is none of these. Whatever its merits, nuclear power is a reality in our society and it is surprising how few people have any idea as to what it is and how it operates, including those who vehemently enter one side or the other of the continuing debate. One area of contention, and misunderstanding, is that which centres around the deliberate introduction into the environment of very low level radioactive wastes. In fact it often comes as a surprise that such deliberate acts are practised at all. This book is an attempt to describe briefly the nature of radioactivity, how it arises in the day-to-day operation of a nuclear reactor, how and why a small fraction is introduced into the environment in a controlled manner, and on what basis such judgements are made. It is an account of current practice, and an attempt to be didactic rather than polemic. As a general subject area it is one which involves many disciplines and therein lies its particular scientific interest, especially for the ecologist.

It is also a subject area of continuing research and one of increasing public awareness.

By the time this book was written (1978-79) SI units had been firmly established. These new units will be used in the future, but most of the published data, upon which this book is based, are in old units. It was therefore decided to use the new units throughout, with the old units immediately following in parentheses. The definitions of both old and new units are given in chapter 1. Indeed the purpose of chapter 1 is simply to introduce some of the basic properties of radioactivity, to comment on its interaction with matter, and to outline some of the methods by which it can be detected. For many readers this may well be unnecessary; and yet for others it may appear to be a rather dull catalogue of obscure definitions, and perhaps too numerate. To the latter I would suggest that the first chapter be read rather quickly, and returned to when necessary for reference. Appendix 4 also gives a list of units and definitions currently used in radiological protection.

Acknowledgements

I would particularly like to thank those colleagues who steered my writings through different subject areas of this book – especially Mr J.R.W. Dutton, Mr D.F. Jefferies and Mr A. Preston – and those who subsequently gave freely of their comments. I am also most grateful to those colleagues who have allowed me to use their published material in the form of tables and figures; due acknowledgement is included in each case. My thanks also to Mrs B.M. Corrigan, who rendered hieroglyphics into type. As for my family, I am not sure which of us suffered the most!

Contents

1. Radioisotopes: their characterization and interaction with matter

1.1. *Introduction*

The generation of power from nuclear energy differs from other sources of energy in that it is based upon the inherent instability of the nuclei of particular heavy atoms which have a tendency, or can be induced, to fall apart. It is the opposite of our principal source of energy, the Sun, in which the energy arises from the fusion of the nuclei of light atoms. The power generated from a nuclear reactor is derived from the binding energy of the nuclei of heavy atoms. Some of the binding energy is converted into kinetic energy of nuclear fragments; this becomes manifest, ultimately, in the form of heat. The heat arises from successive collisions of the fragments with other atoms. Unstable atomic nuclei expend energy in a number of ways, for which the all-embracing term *radiation* is used. Atomic radiations interact with matter in different ways to produce a variety of effects; it is these effects which pose the potential danger of power generation from nuclear energy. But to understand the reasons for this it is first of all necessary to understand something of the nature of the atomic nucleus itself.

1.2. *Atomic species and their behaviour*

All matter is made up of atoms. An atom is composed of a nucleus, which contains almost its entire mass and has a diameter of $\approx 10^{-12}$ cm, surrounded by negatively charged electrons. Including the orbital electrons the atom has a diameter of $\sim 10^{-8}$ cm. There are two principal nuclear particles, or *nucleons*: the positively charged *protons*, and the *neutrons* which are of similar mass to the protons but have no charge. The number of protons in the nucleus (Z) is characteristic of a particular element, but the atoms of an element may have different numbers of neutrons (N) in the nucleus. It is the presence of neutrons that supplies the cohesive force to hold the

1

nucleus together; it would otherwise disintegrate because of the repellent electrical forces of like-charged particles. This cohesive force, the nuclear force, acts over an extremely short range, of the order of 2 to 3 $\times$ 10^{-13} cm. Because the range of nuclear forces is much shorter than the range of the electrical force, it is thought that neutrons can interact only with those nucleons to which they are immediately adjacent, while protons interact with each other even though remotely located within the nucleus. For this reason the number of neutrons must increase more rapidly than the number of protons. The nucleus of the lightest element, hydrogen, contains one proton and the nucleus of the next lightest element, helium, contains two protons and two neutrons. As the number of protons increases, an element may have nuclei with different numbers of neutrons. For example, magnesium has twelve protons but its nuclei may have twelve, thirteen or fourteen neutrons. The *mass numbers* (A) of these three nuclei, i.e. the sum of the protons and neutrons ($A = Z + N$), are therefore 24, 25 and 26, respectively; represented symbolically as ^{24}Mg, ^{25}Mg and ^{26}Mg. These three species of the same element are called *isotopes*. The virtually synonymous term *nuclide* is used to describe a distinct nuclear species; that is an atom characterized by its mass number, its atomic number and the energy state of its nucleus. Isotopes of the same element cannot be distinguished chemically because they have the same electronic structure, the number of electrons – which determines the atom's chemical behaviour – being equal only to the number of protons in the nucleus. Isotopes of the same element are not equally abundant. In the case of magnesium, the majority of naturally occurring magnesium is ^{24}Mg, at 78·70% by weight; the next abundant is ^{26}Mg, at 11·17% and the remaining 10·13% is ^{25}Mg.

As the atomic number (Z) increases, the number of neutrons increasingly exceeds the number of protons. At a Z number greater than 83 the nuclei become unstable. A plot of A number against Z number is shown in figure 1.1. Unstable nuclei disintegrate at a characteristic rate of decay and because this disintegration is accompanied by the emission of various kinds of radiation these unstable isotopes are called *radioisotopes* or, more specifically in discussing one atom in particular, *radionuclides*. It is, in fact, surprising that nuclei are unstable when an excess of neutrons is present. This has been explained in terms of the energy levels of the nucleons grouped on a pairs basis. For example, where nuclei have filled neutron energy levels, but unfilled proton energy levels, the nucleus may undergo an internal rearrangement whereby a neutron

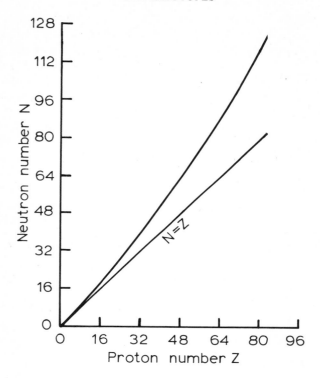

Figure 1.1. Relationship between neutron number (N) and proton number (Z) for stable nuclei.

transforms itself into a proton by emitting an electron. The newly created proton then pairs off with another proton at an unfilled energy level. The exact type of radioactive decay depends upon the type of nuclear instability – if the ratio of protons to neutrons is either too high or too low – and upon the mass – energy relationship between the parent nucleus, the daughter nucleus which results from the parent's decay, and the type of particle emitted.

In order to achieve nuclear stability, the nuclei of radionuclides may emit a variety of particles. Where the ratio of neutrons to protons is too high it appears that, as we have seen, a neutron (n) is converted into a proton (p+) and an electron (e−), the latter being emitted at high speed and having a negative charge:

$$n \rightarrow p^+ + e^-$$

This electron is called a *beta* particle (β^-), and its emission from the nucleus is accompanied by another particle, the *neutrino*, which has no electrical charge and a vanishingly small mass.

Where the ratio of neutrons to protons is too low, an *alpha particle* (*α*) may be emitted. This is a doubly positively charged, massive particle containing two neutrons and two protons, emitted at high speed from the nucleus. With the exception of an isotope of samarium, ^{147}Sm, naturally occurring alpha-emitters have atomic numbers greater than 82. This is because in heavy nuclei the electric forces of the protons, which repel one another, increase more rapidly with increased atomic number than does the cohesive nuclear force. In addition, an emitted particle must also have sufficient energy to overcome the potential barrier at the surface of the nucleus which results from the presence of a large number of positively charged nucleons.

There are a number of nuclei for which the neutron-to-proton ratio is too low but alpha particle emission is not energetically possible. Under certain conditions these nuclei may attain stability by emitting a particle like a beta particle but of opposite charge (e^+). This is a *positron* (β^+) and has only a transitory existence; it combines with an electron and the two particles are annihilated giving rise to two *gamma-ray photons*. The positron, in fact, is thought not to exist independently and is believed to result from the transformation of a proton into a neutron:

$$p^+ \rightarrow n + e^+$$

The loss of a proton from the nucleus leaves the resultant atom in a state of imbalance because it must now lose one of its orbital electrons, the number of which must balance the number of protons.

A neutron-deficient nucleus may also attain stability by quite a different process, the capture of one of its own orbital electrons. The orbital electrons are conveniently visualized as lying in different orbits, or 'shells', around the nucleus, the closest being termed the K shell. Being closest to the nucleus, electrons in the K shell have a greater probability of capture by the nucleus and the process is therefore simply called *K-capture*. The captured electron unites with a proton in the nucleus to form a neutron:

$$p^+ + e^- \rightarrow n$$

The only particle emitted is again the diminutive neutrino. The atom which has suffered the capture of an orbital electron always emits X-rays (an electromagnetic form of radiated energy) which are characteristic of the daughter element. These emissions result from electrons of outer, relatively higher energy, orbits taking up positions of lower energy such that eventually the gap originally occupied by the captured electron is filled.

Some nuclei which have been disturbed, such as by the emission of a subatomic particle, may rearrange themselves to get rid of surplus energy – excitation energy. When this occurs the excess energy is emitted as a *gamma ray* (γ), which is a unit of electromagnetic energy similar to radio waves, or to visible light, but of much shorter wavelength. Indeed it is often convenient to consider gamma rays as *photons*, small packets of energy travelling at the speed of light. Gamma rays have properties indistinguishable from those of X-rays except that they are of nuclear origin and have well-defined wavelengths characteristic of the emitting nucleus. An excited nucleus may alternatively rid itself of its excitation energy by a process known as *internal conversion*, whereby a gamma ray photon transfers its energy to an orbital electron. The electron is ejected from the atom with kinetic energy equivalent to the difference between the energy of the gamma ray photon and the binding energy of the converted electron. After internal conversions, characteristic X-rays are emitted as the outer electrons fill the vacancies left by the converted electrons. These characteristic X-rays may themselves be absorbed, by an internal photoelectric effect of the same nature as internal conversion, by electrons. The electrons ejected by this process are called *Auger electrons*: they possess very little kinetic energy.

1.3. *Radioactive decay*

As a result of the behaviour of unstable nuclei, in which the number of protons is altered, the resultant nuclei are nuclei of different elements. In view of their methods of formation discussed above, the effects of these transformations may be summarized as follows:

(a) *α particle emission* – daughter element has an atomic number two less than the parent, and an atomic mass number four less than the parent.

(b) *β⁻ particle emission* – daughter element has atomic number one higher than that of the parent but has the same atomic mass number as the parent.

(c) *β⁺ particle emission* – daughter element has atomic number one less than that of the parent but has the same atomic mass number as the parent.

(d) *K-capture* – daughter element has atomic number one less than that of the parent but has the same atomic mass number as the parent.

All of the nuclei of a particular radionuclide may not decay by the same mechanism. For example, an isotope of sodium, ^{22}Na, may disintegrate in either of two ways: by the emission of a positron or by K-capture. Positrons are emitted in 89·8% and K-capture occurs in 10·2% of the nuclear transformations. Both methods of decay result in the formation of the same daughter, ^{22}Ne, which is in an excited state. The excitation energy is immediately lost by the emission of a gamma photon. Daughters produced by decay are frequently unstable themselves and may, in turn, undergo nuclear transformations to isotopes of other elements. This process, known as *serial decay*, is an important consideration in studying the potential effects of a particular radionuclide.

The radioactive decay of an atomic nucleus is a completely random event: the probability of an individual unstable nucleus disintegrating at any one time is independent of the fate of neighbouring atoms, the chemical state of the atom – a function of the electrons, not of the nucleus – and of physical conditions. There is, however, a constant probability that the nucleus of a particular radionuclide will disintegrate, denoted by the *decay constant*, λ. Each radionuclide has its own rate of decay and if the number of radioactive atoms of a radionuclide present at time t_0 is N, at some future time (t) there will be fewer unstable atoms of that radionuclide present. The form of the decrease is exponential (figure 1.2.) and may be described as

$$\frac{dN}{dt} = -\lambda N \tag{1.1}$$

where dN/dt is the rate of disintegration, or decay. The result of integrating this equation, and making $N = N_0$ when $t = 0$ is

$$N = N_0 e^{-\lambda t} \tag{1.2}$$

The exact number of radioactive nuclei present at any one time cannot be measured directly. What is measured is the rate of disintegration, i.e. dN/dt, and this is termed *activity* (A). Therefore

$$A = \frac{dN}{dt} = -\lambda N \tag{1.3}$$

and the activity at time t, relative to that at t_0, may be described by

$$A = A_0 e^{-\lambda t} \tag{1.4}$$

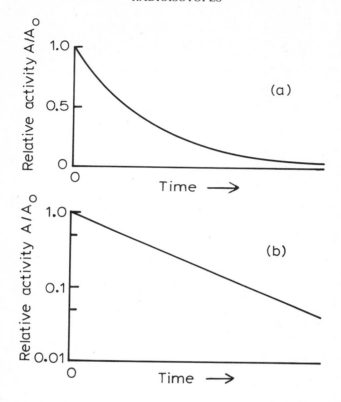

Figure 1.2. Exponential decay of a radionuclide: (a) linear plot; (b) log/linear plot. A is activity at time t, and A_o activity at t_o.

A useful measurement is the time required for any given quantity of radionuclide to decay to one-half of its original amount. This period is called the half-life ($t_\frac{1}{2}$) and may be determined from its rate of decay, as described by equation (1.4), by putting $A = A_o/2$ when $t = t_\frac{1}{2}$. It follows that

$$- \lambda t_\frac{1}{2} = \log_e (\tfrac{1}{2}) \tag{1.5}$$

and therefore

$$t_\frac{1}{2} = \frac{\log_e 2}{\lambda} = \frac{0 \cdot 693}{\lambda} \tag{1.6}$$

Although the half-life of a radionuclide is a unique and repro-ducible characteristic of that radionuclide, it is nevertheless a statistical property and dependent upon the very large numbers of atoms involved. For some applications, therefore, it is convenient to use the *average life* of the radionuclide (τ), defined as the sum of the life-times of the individual atoms divided by the total number of atoms originally present. That is,

$$\tau = \frac{1}{N_o} \int_o^\infty t \, \lambda N_o e^{-\lambda t} dt \tag{1.7}$$

which, upon integration by parts, shows that

$$\tau = \frac{1}{\lambda} \tag{1.8}$$

and therefore

$$\tau = \frac{1}{\log_e 2} t_{\frac{1}{2}} = 1 \cdot 44 \, t_{\frac{1}{2}} \tag{1.9}$$

The unit for the quantity of radioactivity present in a sample must be based on its activity, i.e. on the number of dis-integrations actually occurring. Until recently this unit has been the *curie* (Ci), the number of disintegrations per second of 1 g radium, in fact pure ^{226}Ra. This was defined as

$$1 \text{ Ci} = 3 \cdot 7 \times 10^{10} \text{ disintegrations s}^{-1} \tag{1.10}$$

It is not always, of course, synonymous with the number of particles emitted. For a 'pure' beta emitter, 1 Ci does, in fact, result in the emission of $3.7 \times 10^{10} \beta^-$ particles per second. But for a more complex radionuclide, such as ^{42}K, 20% of the β^- decays are accompanied by a single unit, or quantum, of gamma radiation. The total number of emissions from 1 Ci of ^{42}K is therefore

$$(1 + 0.2) \times 3.7 \times 10^{10} = 4.44 \times 10^{10} \text{s}^{-1} \tag{1.11}$$

The curie is rather a large unit for describing levels of radioactivity in environmental materials, and at laboratory level. A number of sub-multiples are therefore used: for example 10^{-6} Ci is 1 microcurie (μCi) and 10^{-12} Ci is 1 picocurie (pCi). (A full table of prefixes is given in appendix 1, and a list of some radionuclides discussed in the book is given in appendix 2). More recently, a new (SI) unit of radioactivity has been adopted, the *becquerel* (Bq), defined as one disintegration per second. The scale of this unit is thus completely different from the Ci and unfortunately the two are not readily converted one to another. Clearly

$$1 \text{ Ci} = 3.7 \times 10^{10} \text{ Bq} \tag{1.12}$$

but more useful conversion factors to note are that:

$$1 \text{ Bq} = 27.03 \text{ pCi} \qquad (1.13)$$

and that

$$1 \text{ pCi} = 0.037 \text{ Bq (or 37 mBq)} \qquad (1.14)$$

The amount of a radionuclide required to produce a Ci, or a Bq, obviously depends upon both the half-life of the radionuclide and its atomic weight. Two extremes, for example, are ^{238}U which has a half-life of 4.5×10^9 years and 1 Ci of which weighs 2.99×10^6 g; and ^{131}I which has a half-life of only 8 days and 1 Ci of which weighs only 8×10^{-6} g. It can be seen that the amount of an element represented by sub-multiples of a Ci can therefore be extremely small: 1 pCi (37 mBq) of activity is produced by only 8×10^{-18} g of ^{131}I.

The amount of a radionuclide in a sample can also be expressed relative to the total amount of the element present. The two or more isotopes of the same element, both stable and unstable, will behave similarly as long as they are present in identical chemical forms because chemical behaviour is dependent upon the electrons, not the nucleus, of an atom. The amount of radionuclide present, relative to the total weight of element present, is *commonly* known as its *specific activity*; for example as Bq μg^{-1}.[†] It should be stressed, however, that in the environment it is possible that the radionuclide present may, because of its origins, be in an entirely different chemical form from that of the other isotopes of that element. This has important radioecological implications.

It is equally important, in discussing radioactive decay, to note that the amounts of radioactivity of the parent radionuclide and the daughter nuclide, if the latter is unstable, are related one to another with respect to their half-lives. The details of such relationships are given in appendix 3.

1.4. *The interaction of radiation with matter*

The subatomic particles emitted by radionuclides move at extremely high speeds. The kinetic energy of a particle depends upon its mass and a particle travelling with a velocity v, much less than the speed of light, and of mass m, has kinetic energy (E_k) of

$$E_k = \tfrac{1}{2} m v^2 \qquad (1.15)$$

[†] To be precise, specific activity *actually* states the quantity of an individual radionuclide in radiometric terms relative to its quantity in gravimetric terms, i.e. as Bq of isotope per unit mass of that isotope.

Thus a small particle requires a much higher velocity than a large particle to have the same kinetic energy. All electromagnetic radiation travels at the speed of light. Whilst it is essentially a wave motion it also has something of the nature of particulate radiation because it is made of discrete amounts – quanta or photons – of energy. The energy is inversely proportional to the wavelength of the radiation.

The unit used to express the energy of radiation is the *electron volt* (eV). This is equal to the kinetic energy acquired by an electron upon being accelerated through a potential difference of one volt. It is a very small unit,

$$1 \text{ eV} = 1 \cdot 6 \times 10^{-19} \text{ J} \qquad (1.16)$$

Radiation energies are therefore usually expressed as kiloelectron volts (1 keV) = 10^3 eV, and megaelectron volts (1 MeV) = 10^6 eV.

Collisions between emitted charged particles and the nuclei of other atoms are rare; their effects on matter arise from their interaction with the electric fields of the atoms. When a charged particle passes close to an atom there may be a transfer of energy from the particle to an orbital electron. The orbital electron is thereby promoted to a higher energy level within the atom from which it rapidly drops back to its initial ground state, releasing the momentarily acquired energy as a photon. This process is known as *excitation* and the electromagnetic radiation emitted depends upon the energy relationships within the atom whose electron has been excited. The orbital electron may acquire sufficient energy to leave the atom completely. The atom, stripped of an electron, is thereby left with a positive charge. An ion pair is therefore formed – positive atom plus negative electron – and the process is called *ionization*. Sometimes only a single ion pair is formed, sometimes several ion pairs may result; the electron may even acquire sufficient energy to cause further ionizations itself. The kinetic energy of such an electron may be of the order of 1 keV.

A particle gives up energy in producing ionization, the loss of energy being more dependent upon the nature of the medium through which it is passing than upon the initial energy or charge of the particle. Nevertheless, the rate of energy loss does increase as the square of its charge, and is also inversely proportional to its velocity. A fast-moving electron in air loses ~34 eV energy for each ion pair produced. A high-energy particle, such as a 1·7 MeV beta particle, can therefore produce some 5×10^4 primary ion pairs in air.

Beta particles (β^-) are easily deflected. When a beta particle passes close to an atomic nucleus – particularly in a medium of high atomic number – the attractive coulomb force causes the beta particle to deviate; the radial acceleration results in a loss of energy by electromagnetic radiation in the X-ray range. These X-rays are known as *bremsstrahlung*.

The distance which a particle will travel therefore depends upon both its initial energy and the number of ion pairs it forms per unit length; the latter is termed *specific ionization*. Of particular interest, however, is the complementary information of the linear rate of energy absorption as the ionizing particle passes through the medium. This, the *linear energy transfer* (LET), is defined by

$$\text{LET} = \frac{dE_L}{dl} \tag{1.17}$$

where dE_L is the average energy locally imparted to the medium by a charged particle of given energy traversing a distance dl. The LET is usually expressed in units of keV μm^{-1}.

The kinetic energy of an alpha particle is provided by a definite nuclear energy level change of the parent and daughter nuclei involved in the disintegration, so that a particular disintegration gives an alpha particle of characteristic energy. Because they are doubly charged, and of high mass, they create dense tracks of ionization and excitation. As the alpha particle loses its energy its velocity decreases. Ultimately it acquires two electrons from the surrounding medium and becomes a helium atom. Alpha particles have a high linear rate of energy loss and therefore a very short range, the path being in the form of a straight

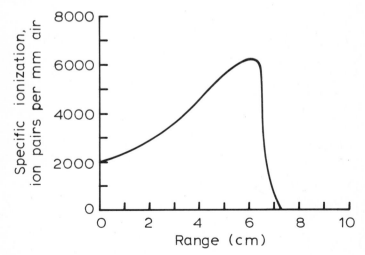

Figure 1.3.   Specific ionization of an α particle.

line. Their range in air is several centimetres, and for energies <4 MeV is given, approximately, as:

$$\text{range (cm)} = 0.56 \, \text{energy (MeV)} \qquad (1.18)$$

In denser media, such as water, their range is only a few tens of micrometres. The number of ion pairs formed per unit length increases as the alpha particle slows down (figure 1.3.).

Beta particles also cause excitations and ionizations but, being of much smaller mass, and having only half the charge of an alpha particle, they are more readily scattered and penetrate more deeply into the medium. They therefore have a lower specific ionization. Beta particles are not mono-energetic but have a continuous energy spectrum up to a maximum for a particular radionuclide. The maximum energy for pure beta emitters is approximately equivalent to the difference in mass between the parent nucleus and that of the daughter nucleus plus beta particle. It might be expected, therefore, that the beta particles should be mono-energetic, but in fact the accompanying neutrino, the simultaneous production of which was mentioned previously, shares the available energy. A typical beta spectrum is shown in figure 1.4.

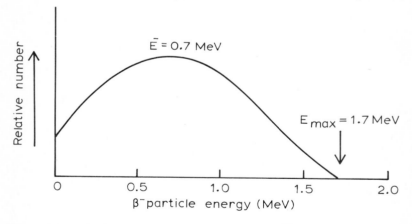

Figure 1.4.   The β^-particle energy spectrum of ^{32}P. $\bar{E}$ is the energy at which there is the greatest probability of a beta emission, and E_{max} corresponds to the beta particle taking all of the energy of the decay.

The average energy ($\bar{E}$) of a beta particle is between 30% and 40% of the maximum energy (E_{max}). The maximum range of beta particles is dependent both upon their maximum energies (figure 1.5), and upon the density of the medium. The number of ion pairs formed per unit length, however, is relatively high for low-energy beta particles; it then

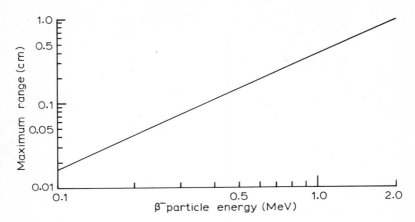

Figure 1.5.   Range of β^-particles in water as a function of their energy.

decreases as the energy of the beta particle increases until, at about
1 MeV, a broad minimum is reached (figure 1.6.). At higher beta
energies there is a slight increase. The number of ion pairs formed per
unit length of track for a beta particle of a particular energy is also not
constant. Emitted from the nucleus at close to the speed of light, a
high-energy beta particle loses initially very little energy; but at the end
of its track its specific ionization will rapidly increase to about 300

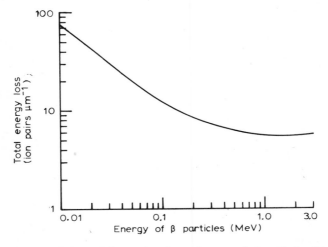

Figure 1.6.   Number of ion pairs formed per unit length by β^-particles,
travelling through water, as a function of their energy.

times its initial value. With high-energy beta particles in media of high atomic number the accompanying bremsstrahlung are also an important consideration. Positrons have a similar penetration to beta particles. As soon as their kinetic energy has been expended, however, they are immediately attracted to the nearest electron, the two are annihilated, and the resultant energy is almost always emitted as two gamma rays each of 0·51 MeV.

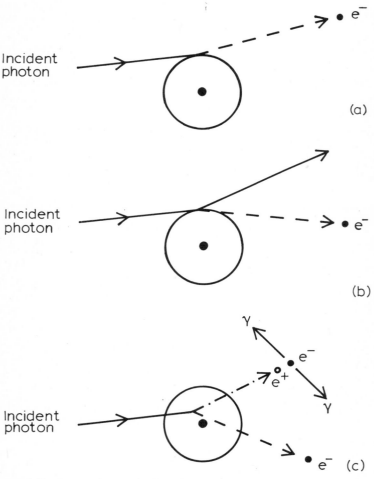

Figure 1.7. Interactions of electromagnetic radiations with matter: (a) photo-electric absorption; (b) Compton scattering; (c) pair production – each γ-ray is 0·51 MeV.

Beta particles are scattered in absorbing media and their direction may be completely reversed. This is known as *backscatter*. Notwithstanding backscatter, it will be clear that because of the limited energy ranges of both beta and alpha particles it is possible that many of them could be absorbed by the source material itself. This self-absorption is particularly important when trying to estimate the quantity of alpha and beta emitting nuclides in, for example, environmental materials.

Electromagnetic radiations – gamma rays and X-rays – are uncharged and have no rest mass. They cannot, therefore, cause ionization of atoms by direct interaction with the atomic force field. Electromagnetic radiations can react with matter in a number of different ways, however, three of which result in the production of secondary electrons which in turn cause ionizations in a manner similar to beta particles. A photon may be completely absorbed by an interaction with a tightly bound electron, the binding energy of which is equal to, or less than, the energy of the photon. Virtually all of the energy of the photon is transferred to the electron, which is ejected from its orbital. This process (figure 1.7 (a)) is therefore called *photoelectric absorption* and predominates at low gamma energies and in absorbing media of high atomic number. At intermediate energies, of >0·5 MeV, an elastic collision may occur between a photon and an electron which has a lower binding energy to its nucleus than that of the photon. It is impossible for all of the photon's energy to be transferred to the electron and it is therefore scattered possessing less energy – a longer wavelength – than the incident photon. The energy difference is transferred to the orbital electron which is therefore also scattered. This process (figure 1.7 (b)) is known as *Compton scatter*, or *Compton recoil*.

A third possibility, for gamma rays whose energy exceeds 1·02 MeV, is that as they pass close to the nucleus of an atom they vanish, the gamma ray's energy reappearing as a positron and an electron, which have a combined rest mass of 1·02 MeV. All energy in excess of 1·02 MeV is imparted to the created particles as kinetic energy. This transformation of energy into mass needs an intensely strong field, such as is found near a nucleus. The electron and positron lose their kinetic energy by excitation, ionization and bremsstrahlung; the positron, having expended its kinetic energy, unites with an electron and, by annihilation, forms two gamma-ray photons of 0·51 MeV each. This event (figure 1.7 (c)) is known as *pair production* and occurs most frequently when high-energy gamma-rays interact with elements of low atomic number.

The absorption of gamma-rays is qualitatively different from that of particle radiation. The latter have finite ranges in matter whereas gamma rays do not. Gamma radiation is much more penetrating than particle radiation but it has a very low specific ionization, ~1.5 cm^{-1} in air. The rate at which gamma radiation is absorbed in a given material is specific for any energy level, given than the gamma rays are presented as a narrow beam. The *linear attenuation coefficient* (μ) can be calculated from

$$\frac{I}{I_0} = e^{-\mu x} \qquad (1.19)$$

where I is the intensity of radiation after passing through a given thickness (x) of material, and I_0 the initial intensity of the radiation at zero thickness. When the absorbing material is extensive in all directions the radiation is more greatly scattered; some radiation will even be scattered back into the initial radiation beam, thereby increasing the radiation intensity at a given point. This increase is expressed by a build-up factor which can be used to modify the simple equation (1.19) above.

Neutrons, as we have seen, are present in the atomic nucleus and have no charge. Free neutrons are produced when alpha or gamma radiations interact with light elements. A typical reaction is that of an alpha particle and beryllium (^{9}Be) which results in a compound nucleus of carbon being formed, ^{13}C, in an excited state. The compound nucleus immediately rids itself of the excitation energy by releasing one neutron of high energy, becoming ^{12}C. Another method of neutron formation is nuclear fission. A number of very heavy nuclei decay by *spontaneous fission* in which the parent nucleus breaks very approximately in half and releases a number of neutrons: the fission can also be *induced*, as will be discussed in chapter 4. Neutrons with high energies, in excess of 0.1 MeV, are termed *fast* neutrons, and those with very low energies (~ 0.025 eV) are termed *thermal neutrons* because they are in approximate thermal equilibrium with their surroundings. Thermal neutrons are obtained from fast neutrons by deliberately slowing them down, a process called *moderation*.

Because neutrons have no charge they interact with atomic nuclei after being slowed down by colliding elastically or inelastically with surrounding atoms. In elastic collisions with nuclei the neutron loses some of its initial energy, which is transferred to the target nucleus. A head-on collision with a proton may result in almost total energy transfer. The transferred energy appears as kinetic energy of the target nucleus. Only some of the kinetic energy may be given up to the nucleus, however, resulting in inelastic collisions whereby the nucleus becomes excited, the excitation energy being emitted as a gamma-photon. Both processes result in varying degrees of ionization, but the former is obviously very effective. Slow neutrons may become incorporated into the nuclei of atoms, which therefore increase by one mass number. This process is called *neutron activation* and is a means of artificially producing radioisotopes for industrial, scientific and medical purposes. Following such absorption of a neutron, the nucleus may react by either emitting a sub-atomic particle or shedding excitation energy in the form of gamma-photons. An excited nucleus of a specific isotope may thus decay in a number of ways; for example $^{36}Cl^*$ may decay into ^{36}Cl, ^{35}Cl, ^{34}Cl, ^{35}S or ^{32}P, each nuclide having its own probability of formation.

The distance which neutrons can travel is very much greater than that of charged particles because they are not affected by electrons, and nuclei occupy such a small fraction of the volume of an atom. It follows, therefore, that a medium such as water – which has about 30 times as many nuclei per unit mass as lead – will be much more effective in stopping neutrons. The converse holds for charged particles and gamma radiation, for which dense materials such as lead provide the best shielding. 'Free' neutrons are not stable but decay to form a $p + e^-$.

The interactions of radiation with matter, as briefly described above, are similar for both inorganic and organic materials: they take place in a very short space of time ($\sim 10^{-16}$ s). When these reactions – which produce ionizations and excitations – occur in biological materials the resultant chemical damage may result in ultimate biological damage. The complex nature of this process is the subject of another book in this series (No. 14; *Biological Effects of Radiation*, by J.E. Coggle) and only very brief mention will be made of it here. Biologically important molecules may be damaged in two ways: by the direct deposition of energy in a molecule as a result of its primary interaction with the radiation, or by an indirect action of the radiation on water – the principal component of biological tissues – as a result of radiolysis by which highly reactive products are produced. These products may react with organic biological molecules or, of particular importance, with oxygen. The first step is the formation of ionized and excited molecules. One ionization event requires the transfer of ~ 34 eV energy, although it is considered that most primary ionization events require the transfer of ~ 100 eV energy in a very localized area. Lower transfers of energy result in the excitation of molecules.

Probably the most important indirect effect of radiation is the formation of *free radicals* – atoms or fragments of a compound that contain an unpaired electron. This may be preceded by ionization. For example

$$A \rightarrow A^+ + e^-$$

The electron is quickly captured, according to the relative electron affinities of surrounding molecules (B),

$$e^- + B \rightarrow B^-$$

For example the effect on water may be considered as

$$H_2O \rightarrow H_2O^+ + e^-$$

followed by

$$e^- + H_2O \rightarrow H_2O^-$$

Each H_2O ion subsequently forms an ion plus a free radical (denoted by a dot) such that

$$H_2O^+ \rightarrow H^+ + OH \bullet$$
$$H_2O^- \rightarrow OH^- + H \bullet$$

The product may be water plus $H \bullet$ and $OH \bullet$, the free radicals reacting in a number of ways, either with each other or with other molecules, in a chain of damaging reactions. Particularly damaging is the direct production, from the combination of two $OH \bullet$ radicals, of hydrogen peroxide (H_2O_2), an active oxidizing agent. It may also be formed, via the hydroperoxy radical ($HO_2 \bullet$), either by combination or by the capture of an electron followed by a hydrogen ion. Aqueous free radicals may react with organic molecules (represented by RH) by, for example

$$RH + OH \bullet \rightarrow R \bullet + H_2O$$
$$\text{or} \quad RH + H \bullet \rightarrow R \bullet + H_2$$

The net result of such effects, of which the above are the merest sample, is a wide range of subsequent biological damage. The amount and type of resultant biological damage is, however, largely dependent upon the amount of radiation received by the biological material, the rate at which it is received, and the type and nature of the receiving material. It is also therefore necessary to discuss measurements of radiation with regard to the amounts given and received.

1.5. Radiation dosimetry

The units of exposure to radiation are defined in physical terms to which, where applicable, biological modifying factors are added. The original unit used was the *röntgen*, also spelt *roentgen*, abbreviated to R. This unit expresses the quantity of ionization produced by radiation in air, or the capacity of radiation to ionize air. It has a precise definition: the röntgen is that quantity of X or gamma-radiation the associated corpuscular emissions of which per 0.001 293 g (1 cm^3) of dry air produce ions carrying 1 electrostatic unit of electric charge of either sign. In effect this results in 2.1×10^9 ion pairs cm^{-3} being produced by the penetration of air by 1 R. By definition, therefore, the röntgen is limited to describing electromagnetic radiation – in fact below 3 MeV – and to air. It should also be noted that it is a unit of integrated exposure and is independent of the time over which the exposure occurs. The total exposure is the product of exposure rate (usually expressed as R h^{-1}, or mR min^{-1}) and time.

A much more useful unit is one of absorbed dose (D), for as we have seen the interaction of radiation with matter depends upon both the type of radiation and the material with which it interacts. The unit for radiation absorbed dose has, until recently, been an acronym of it, the *rad*. This has been defined as an energy absorption of 100 erg g^{-1} (0.01 J kg^{-1}) of a specified material from *any* ionizing radiation. For gamma radiation in soft tissue the absorbed dose per röntgen is approximately 1 rad.

A new unit has recently been introduced which, like the becquerel replacing the curie, will replace the rad. This new unit is the *gray* (Gy), defined as an energy absorption of 1 J kg^{-1}. Therefore

$$1 \text{ Gy} = 100 \text{ rad} \qquad (1.20)$$

These units are of limited value for defining the relative effects of the interaction of radiation with biological materials. As discussed above, the amount of energy transferred per unit length (LET) differs markedly for different types of radiation and, because of its interaction with water and organic molecules, the higher the LET the more effective is the damage likely to be to the organism. Therefore radiation dissipating 1 rad with a high specific ionization will have a greater biological effect than will radiation dissipating 1 rad with a low specific ionization. In order to provide a standard of comparison for the combination of the biological effects of different types of radiation, a further unit has been derived to describe this *relative biological effectiveness*, the R.B.E. Based on a comparison with a standard, the R.B.E. is defined as

$$\text{R.B.E.} = \frac{\text{dose (e.g. in Gy) from 200 keV X-rays causing a specific effect}}{\text{dose (e.g. in Gy) from radiation causing the same effect}} \quad (1.21)$$

The concept of R.B.E. has been shown to be capable of comparing observable biological effects of different types of radiation, and it is still employed in radiation biology. As a modifying factor in assessing biological effects in the field of radiological protection, however, even this concept is rather limited because the dose levels being considered are such that there is only a very small probability of there being biological effects to observe. In addition, at very low dose rates, there is the complication that a number of different biological effects would have a low probability of occurring. Nevertheless, it is necessary to have a means of quantifying the relative effects of different radiations on man, and this normalizing quantity is called the *quality factor* (Q). The quality factor relates to different values of LET as given in table 1.1 and, for convenience, relates approximately as the *effective quality factor* $(\bar{Q})$, to various types of radiation as given in table 1.2.

TABLE 1.1. *Relationship between quality factor (Q) and linear energy transfer (LET).*

LET (keV μm^{-1} in water)	Q
$<$ 3·5	1
3·5– 7·0	1– 2
7·0– 23·0	2– 5
23·0– 53·0	5–10
53·0–175·0	10–20
$>$175·0	20

TABLE 1.2. *Values for the effective quality factor ($\bar{Q}$) related to the various types of primary radiation.*

Radiation	$\bar{Q}$
X-rays, γ-rays and electrons	1
Neutrons, protons and singly-charged particles of rest mass of greater than one atomic mass unit of unknown energy	10
α particles and multiply charged particles (and particles of unknown charge), of unknown energy	20

We therefore have units with which to measure exposure to radiation, the absorption of radiation, and some of the parameters which affect the latter; these are still insufficient for radiation safety purposes. One wishes to consider not only the relative biological effect of the radiations from a particular radionuclide but also to allow for other factors such as its varying distribution within a tissue or organ of the body, or the absorption of different radiations by the whole body. The unit used to describe the product of these different factors, that is the absorbed dose (D) in rads multiplied by the quality factor (Q) and by any other modifying factor, has been the *rem*. Because the rem is based upon the rad, this unit too has recently been changed. The SI unit is the *sievert* (Sv):

$$1\,Sv = 100\,rem \qquad (1.22)$$

The *dose equivalent*, H, in rems or sieverts, expresses the *total* biologically effective dose regardless of the type of radiation. It also enables the summation of a mixture of different types of radiation from different, simultaneous, sources. By a slight modification it can also be used to state an estimate of dose to a population residing in a given area, or for the total population of a country, by multiplying the average dose value by the number of people. This, the *collective dose equivalent*, (S) is sometimes expressed as a *man-sievert* or *man-rem* unit. It is particularly valuable because it can represent real people, the number of which has been measured and their dose rates and other parameters actually studied.

1.6. *Radiation detection and measurement*

It is a paradox that one of the reasons for public concern about radioactivity is that it cannot be detected by any of the five senses of man; and yet of all the different classes of chemicals which have the potential to contaminate or pollute the environment radionuclides are the easiest to detect instrumentally – both in terms of the very low quantities which can be detected and the speed with which such analyses can, generally, be made. The methods of radiation detection are based mainly on the ability of radiation to cause ionization of gases, excitation in liquids and solids, and chemical changes. Three types of analysis need to be made: the quantitative detection and characterization of radionuclides in a sample, the dose rate resulting from their presence, and the total dose received by an object in their presence.

A quantitative assessment of total alpha or beta radiation present is made by *gas ionization counting* methods. These methods rely on the production of ion pairs by particle radiations in gases. The ion pairs allow an electric current to flow if a voltage is applied across the gas chamber. At higher applied voltages the formed electrons are accelerated to cause further ionizations and thus amplify the original ionizing event; in this manner the number of ion pairs can be amplified by a factor of ~10^8. The gas chamber may have an end window, against which the sample is presented, or be window-less in which case the sample is placed at the open end and a gas made to flow through the chamber. The gas chambers are connected to a variety of electronic apparatus which produce numerical data for the unknown sample which can be compared with known standards. Both total alpha and total beta measurements are complicated because of the absorption of emitted particles by the source material itself. Samples are therefore usually ashed, to reduce their bulk, and presented to the counters in as thin a source as possible.

Gas ionization chambers can be refined to provide some degree of characterization of radiation emitted; but they have a very low efficiency for gamma radiations, which have a low specific ionization. Gamma radiations, and more detailed analyses of particle radiations, are therefore made using *scintillation counting* techniques. Scintillation counting utilizes the property of radiations to cause excitation in matter, as described above, and enables more dense materials to be used. The energy of an excited orbital electron, relative to its ground state, varies from one substance to another. The amount of excitation energy released therefore also varies, as does the wavelength, which is inversely related to energy. For some materials – called *scintillators* or *fluors* – the energy relationships are such that the wavelength of electromagnetic radiation produced, as a result of secondary electrons activated within the fluor, are in the visible or near u.v. range. The resultant scintillations are detected by a photocell which converts the photons to electrons, the electrons are then amplified in number by photomultipliers and a quantitative relationship established between the excitation energy imparted to the fluor and the voltage of the pulse obtained from the other end of the photomultiplier. These pulses can be energy-discriminated, segregated into channels, and counted to provide a spectrum of number of pulses received plotted against energy. This is usually done automatically and presented in a visual display on a cathode ray tube. An example is shown in figure 1.8.

A typical scintillator for counting gamma radiation is a crystal of sodium iodide containing about 1% thallous iodide. Because gamma rays have well-defined wavelengths, characteristic of the emitting

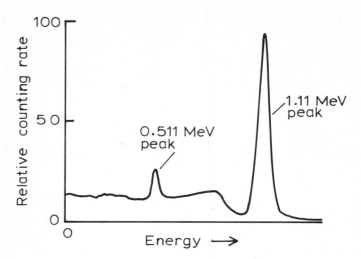

Figure 1.8. Gamma-ray spectrum of ^{65}Zn, showing the 0·511 MeV peak resulting from e$^+$ annihilation.

nucleus, it is possible to identify different nuclides by their characteristic spectra. Spectral resolution varies, depending upon the type of crystal and the number of channels used. When a number of radionuclides are present in a mixture there will be a considerable overlap, and some degree of chemical separation may be required initially. *Gamma-spectrometry* is widely used, with computer methods for resolving complex spectra, in environmental surveillance. Radio-nuclides decaying by alpha particle emission can also be identified because of the characteristic energies of the alpha particles, but it is first of all necessary to separate chemically the element of interest. *Alpha-spectrometry* is made using thin silica crystals against which the sample is presented plated out on to a small planchette.

Beta radiations can also be determined by activated crystal methods but, because of the varying range of energies resulting from beta particle emission, radionuclide identification is extremely difficult. Of course many beta-emitting radionuclides also emit characteristic gamma radiation, but for those which are pure beta emitters – and this group includes a number of biologically important elements – it is more prudent to separate chemically the element from the material, and other radionuclides, and analyse by a *liquid scintillation* method. This technique involves dissolving the sample with the fluor – an organic compound – in the same solution, the emitted photons again being detected by photomultipliers. Alpha-emitters can also be measured in this way. Finally, although many decay processes involve X-ray emission, it is not usually necessary to resort to this form of detection; an important exception is the detection of ^{55}Fe.

Radiation doses may be measured by a number of different instruments. Radiation monitors are based on ionization chambers for beta and gamma radiations, and scintillation detectors for alpha radiations, coupled with an electronic circuit that integrates the radiation measured over a set period of time. The radiation received is expressed as absorbed dose per unit time. Personal dosimetry has most commonly been made by using film badges, which consist of a film covered by a number of filters to give some indication of the different types of radiation which may have exposed it. More recently, thermoluminescent dosimeters (TLDs) have been introduced to replace film badges. Quartz-fibre dosimeters, photoluminescent dosimeters and a variety of other types may also be used. For areas around nuclear reactors neutron monitors are also required. These use elements of high neutron-capture cross-section, such as boron or cadmium, surrounded by a moderator, such as Polythene, to slow down the neutrons. Many of the radiation monitors – and some of the analytical instrumentation – can be taken into the field. It is therefore possible to make on-the-spot assessments of the presence or absence of radioactivity above background levels, the dose which can be received from them, and even some assessment of which radionuclides are present.

2. The radiation background

2.1. *Introduction*

Natural radioactive elements have existed in the Earth since its creation some 4.6×10^9 years ago. All elements of atomic number greater than 80 possess radioactive isotopes, and all isotopes of elements greater than atomic number 83 are radioactive. In addition, the earth is constantly bombarded by cosmic rays. To these natural sources of radiation must be added the fallout resulting from the atmospheric testing of nuclear weapons which, to a varying degree, has increased the exposure to man. The development of diagnostic medical techniques – using ionizing radiations and radiopharmaceuticals – has also contributed to the life-time exposure of man in many areas of the world. In order to assess the impact of nuclear power programmes on man and the environment it is therefore necessary briefly to review this radiation background against which any assessment can be made.

2.2. *Cosmic radiation*

Primary cosmic radiation impinges constantly upon the Earth. It originates mainly within our own galaxy and at times of intense solar activity the contribution from the Sun increases considerably. Primary cosmic radiation consists mainly of very high-energy protons, with alpha particles and the nuclei of a number of elements comprising less than 20% of the total. These primary particles interact with the upper atmosphere to produce a number of secondary particles, such as neutrons, protons, mesons (particles which have a mass intermediate between nucleons and electrons) and gamma radiation. The charged particles are affected by the magnetic field of the Earth; consequently, the amount of both primary and secondary cosmic radiations reaching the Earth's surface varies with latitude and is, in general, lower at the middle latitudes. Recent estimates of the total absorbed dose rate at sea level, out of doors, resulting from this ionizing component is about 0·032 μGy h^{-1} (3·2 μrad h^{-1}); the annual absorbed dose in human tissues is approximately 0·28 mGy (28 mrad). The neutron absorbed dose index rate at sea level is very low, at about 0·0004 μGy h^{-1} (0·04 μrad h^{-1}), giving an annual absorbed dose in human tissues at sea level of between 2 and 3·5 μGy (0·2 and 0·35 mrad), depending upon latitude.

The atmosphere acts as a shield against cosmic radiation and therefore the dose received will increase with altitude. Thus at a latitude of 50° the cosmic ray intensity at about 1500 m is 60% greater than at sea level, and at 3000 m it is more than three times the sea-level value. The relationship between dose rate and altitude is complicated by the fact that the shielding depends upon the *depth* of atmosphere, and this varies with latitude. At sea level this results in the cosmic ray intensity being about 12% greater at the poles than that at the equator, whereas at 3000 m the intensity at the poles is about 50% greater than that at the equator.

The recent development of transport by supersonic aircraft has increased interest in estimating the absorbed dose resulting from high altitude flight. The data given in table 2.1 have been calculated to compare a number of flight paths at either subsonic or supersonic speed. The supersonic aircraft fly at a much greater altitude but their flight duration is much shorter. Consequently, the total absorbed dose is actually less than for subsonic flights. At times of solar flares, however, the situation is very different and dose rates increase by

TABLE 2.1. *Comparison of calculated cosmic-ray doses to a person flying in subsonic and supersonic aircraft*

Route	Subsonic flight at 11 km height			Supersonic flight at 19 km height		
	Flight time (h)	Dose per round trip μGy	(mrad)	Flight time (h)	Dose per round trip μGy	(mrad)
Paris–Los Angeles	11·1	48	(4·8)	3·8	37	(3·7)
Paris–Chicago	8·3	36	(3·6)	2·8	26	(2·6)
Paris–New York	7·4	31	(3·1)	2·6	24	(2·4)
New York–Los Angeles	5·2	19	(1·9)	1·9	13	(1·3)
Acapulco–Sydney	17·4	44	(4·4)	6·2	21	(2·1)

Data from the Report to the United Nations Scientific Committee on the Effects of Atomic Radiation, 1977, General Assembly document 32 Session, Supplement No. 40 (A/32/40) (New York: United Nations).

more than an order of magnitude. Solar flares occur a few times in each solar cycle but usually last only for a few hours. Supersonic aircraft carry radiation monitors and therefore descend to lower altitude flight paths when necessary.

2.3. *Terrestrial radiation*

When the Earth was first formed a relatively large number of isotopes would have been radioactive. Those radionuclides with half-lives of less than about 10^8 years would by now have decayed to undetectable levels. The naturally occurring radionuclides on Earth are therefore those with relatively long half-lives, and these can be divided into those which occur singly and those which are components of three distinct decay series. Of the former the most important is ^{40}K, with a half-life of $1 \cdot 27 \times 10^9$ years, which emits both beta and gamma radiation. Natural potassium consists principally of the stable isotope ^{39}K, and only 0·012% by weight is the radioactive ^{40}K. Another radionuclide of importance is ^{87}Rb, which emits beta radiation, has a half life of $4 \cdot 8 \times 10^{10}$ years, and an isotopic abundance of 27·9%. Both of these radionuclides decay to stable isotopes. In contrast, two radionuclides of uranium, and one of thorium, decay to give rise to families of radionuclides which decay in three distinct series. All three series contain alpha emitters. One begins with the decay of ^{238}U (half-life $4 \cdot 5 \times 10^9$ years) and is called the *uranium series*; a second begins with ^{232}Th (half-life $1 \cdot 4 \times 10^{10}$ years), the *thorium series*; and the third begins with ^{235}U (half-life $7 \cdot 1 \times 10^8$ years), the so-called *actinium series*. All three decay, through three complex series, to stable isotopes of lead; ^{206}Pb, ^{208}Pb and ^{207}Pb respectively. The three series are presented in tables 2.2, 2.3 and 2.4.

To these terrestrial radionuclides must be added those radionuclides which are formed as a result of neutron capture in the atmosphere, notably ^3H (tritium), ^{14}C and ^7Be. The use of ^{14}C in geochronology is well-known. It has a half-life of 5730 years and is produced by the absorption of a neutron by ^{14}N. The number of atoms produced by this process is estimated to be about 2·3 atoms s^{-1} cm^{-2} of the atmosphere, resulting in a global inventory of about 11·1 EBq (0·3 GCi).

The decay of natural radionuclides produces both particle and electromagnetic radiation. Because human organs and tissues are shielded to a large extent from particle radiation, only the gamma contribution is really important in estimating the *external* radiation resulting from these radionuclides. Different types of rock, and the

TABLE 2.2   *Nuclides of the uranium (^{238}U) series.*

Nuclide	Half-life	Type of decay
^{238}U	4.5×10^9 y	α
^{234}Th	24 d	β^-
^{234m}Pa	1.2 m	β^-
^{234}U	2.5×10^5 y	α
^{230}Th	7.7×10^4 y	α
^{226}Ra	1600 y	α
^{222}Rn	3.8 d	α
^{218}Po	3.1 m	α, β^-
^{218}At	~2 s	α, β^-
^{218}Rn	3.0×10^{-2} s	α
^{214}Pb	26.8 m	β^-
^{214}Bi	19.8 m	β^-, α
^{210}Tl	1.3 m	β^-
^{214}Po	1.6×10^{-4} s	α
^{210}Pb	22.3 y	β^-
^{210}Bi	5.0 d	β^-, α
^{206}Tl	4.2 m	β^-
^{210}Po	138 d	α
^{206}Pb	stable	

y = years; d = days; m = minutes; s = seconds.

soils resulting from them, therefore give rise to different dose rates because of their different concentrations of these radionuclides. This is a very important consideration because it immediately demonstrates that the external radiation dose received by man varies not only with altitude but with the very ground upon which he stands, and with the building materials he derives from it. In order to obtain some idea of the range of doses which can be absorbed as a result of radionuclides in some of the more common rocks and soils, table 2.5 gives their

TABLE 2.3. *Nuclides of the thorium (^{232}Th) series.*

Nuclide	Half-life	Type of decay
^{232}Th	$1{\cdot}4 \times 10^{10}$ y	α
↓		
^{228}Ra	$5{\cdot}8$ y	β^-
↓		
^{228}Ac	$6{\cdot}1$ h	β^-
↓		
^{228}Th	$1{\cdot}9$ y	α
↓		
^{224}Ra	$3{\cdot}6$ d	α
↓		
^{220}Rn	55 s	α
↓		
^{216}Po	$0{\cdot}15$ s	α
↓		
^{212}Pb	$10{\cdot}6$ h	β^-
↓		
^{212}Bi	$60{\cdot}6$ m	α, β^-
^{212}Po	$3{\cdot}1 \times 10^{-7}$ s	α
^{208}Tl	$3{\cdot}1$ m	β^-
^{208}Pb	stable	

y = years; d = days; h = hours; m = minutes; s = seconds.

concentrations in these materials, and the estimated absorbed dose rates in air at 1m above the surface. It is immediately apparent that the dose rates can differ by an order of magnitude. The main contributors to the absorbed dose rate in air are the radionuclides ^{208}Tl and ^{228}Ac in the ^{232}Th series, and two short-lived decay products of ^{222}Rn, namely ^{214}Pb and ^{214}Bi in the ^{238}U series. Snow cover and soil water content can markedly affect these dose rate estimates.

The world-wide average external dose rate at 1m above the ground resulting from terrestrial radionuclides is taken to be 0·045 μGy h^{-1} (4·5 μrad h^{-1}). There are areas, however, which are known to have exceptionally high dose rates and these are of special interest, particularly with regard to the possible effects of radiation on man. The two best studied areas are parts of Brazil and India. In Brazil for example, there are two distinct types of deposit high in radioactive content; the black monazite sands – which are high in thorium series radionuclides – of certain beaches in the states of Espirito Santos and Rio de Janeiro, and volcanic regions in the state of Minas Gerais. The monazite sands area, particularly the town of Guarapari, has absorbed

TABLE 2.4. *Nuclides of the actinium (^{235}U) series.*

Nuclide	Half-life	Type of decay
^{235}U	$7 \cdot 1 \times 10^8$ y	α
^{231}Th	$25 \cdot 5$ h	β^-
^{231}Pa	$3 \cdot 3 \times 10^4$ y	α
^{227}Ac	$21 \cdot 8$ y	β^-, α
^{223}Fr	22 m	β^-
^{227}Th	$18 \cdot 5$ d	α
^{223}Ra	$11 \cdot 4$ d	α
^{219}Rn	4 s	α
^{215}Po	$1 \cdot 8 \times 10^{-3}$ s	α, β^-
^{215}At	$1 \cdot 0 \times 10^{-4}$ s	α
^{211}Pb	36 m	β^-
^{211}Bi	$2 \cdot 1$ m	α, β^-
^{211}Po	$0 \cdot 56$ s	α
^{207}Th	$4 \cdot 8$ m	β^-
^{207}Pb	stable	

y = years; d = days; h = hours; m = minutes; s = seconds.

dose rates in air along the streets of 1 to 2 μGy h^{-1} (0·1 to 0·2 mrad h^{-1}). Some areas of the beach – a popular holiday resort – have dose rates of 20 μGy h^{-1} (2 mrad h^{-1}). The inhabited volcanic regions of Minas Gerais have absorbed dose rates in air of up to 4 μGy h^{-1} (0·4 mrad h^{-1}). Such areas are not unique but others have been less well studied. Areas of high natural radiation also occur in Iran and France and locally in a number of countries which have mineral springs.

All of the above absorbed dose rates refer to outdoor measurements. The majority of individuals in countries most likely to have nuclear power programmes spend a large part of their time indoors. It might be imagined, therefore, that inside buildings a certain degree of shielding exists to reduce the dose rate. Building materials, however, also contain the naturally occurring radionuclides. Of the materials commonly used for construction the one which contains the least

TABLE 2.5. *Typical concentrations of ^{40}K, ^{238}U and ^{232}Th in common rocks and soils, and the estimated absorbed dose rates in air at 1 m above the surface.*

Type of rock or soil	Concentration, Bq g⁻¹ (pCi g⁻¹)			Absorbed dose rate in air, μGy h⁻¹ (μrad h⁻¹)
	^{40}K	^{238}U	^{232}Th	
Igneous rock				
Granite	1·00 (27·0)	0·059 (1·60)	0·081 (2·20)	0·120 (12·0)
Diorite	0·70 (19·0)	0·023 (0·62)	0·033 (0·88)	0·062 (6·2)
Basalt	0·24 (6·5)	0·011 (0·31)	0·011 (0·30)	0·023 (2·3)
Durite	0·15 (4·0)	0·0004 (0·01)	0·024 (0·66)	0·023 (2·3)
Sedimentary rock				
Limestone	0·09 (2·4)	0·028 (0·75)	0·007 (0·19)	0·020 (2·0)
Carbonate	—	0·027 (0·72)	0·008 (0·21)	0·017 (1·7)
Sandstone	0·37 (10·0)	0·019 (0·50)	0·011 (0·30)	0·032 (3·2)
Shale	0·70 (19·0)	0·044 (1·20)	0·044 (1·20)	0·079 (7·9)
Soil				
Serozem	0·67 (18·0)	0·031 (0·85)	0·048 (1·30)	0·074 (7·4)
Chernozem	0·41 (11·0)	0·021 (0·58)	0·036 (0·97)	0·051 (5·1)
Sodpodzolic	0·30 (8·1)	0·015 (0·41)	0·022 (0·60)	0·034 (3·4)
Boggy	0·09 (2·4)	0·006 (0·17)	0·006 (0·17)	0·011 (1·1)

Data from the Report to the United Nations Scientific Committee on the Effects of Atomic Radiation, 1977, General Assembly document 32 Session, Supplement No.40 (A/32/40) (New York: United Nations.)

TABLE 2.6. *Concentrations of ⁴⁰K, ²²⁶Ra and ²³²Th in some building materials and the absorbed dose rate in air within a room constructed of them assuming 4 π geometry.*

Type of building material	Country	Concentration, Bq g⁻¹ (pCi g⁻¹)			Absorbed dose rate in air, µGy h⁻¹ (µrad h⁻¹)
		^{40}K	^{226}Ra	^{232}Th	
Bricks	Sweden	0·93 (25)	0·096 (2·6)	0·126 (3·4)	0·33 (33)
	UK	0·63 (17)	0·052 (1·4)	0·044 (1·2)	0·16 (16)
	USSR	0·74 (20)	0·056 (1·5)	0·037 (1·0)	0·16 (16)
Concrete	Sweden	0·70 (19)	0·048 (1·3)	0·085 (2·3)	0·21 (21)
	UK	0·52 (14)	0·074 (2·0)	0·030 (0·8)	0·15 (15)
	USSR	0·56 (15)	0·033 (0·9)	0·030 (0·8)	0·12 (12)
Plaster	Sweden	0·02 (0·6)	0·003 (0·09)	<0·001 (<0·04)	<0·01 (<1)
	UK	0·15 (4·0)	0·022 (0·60)	0·007 (0·20)	0·04 (4)
	USSR	0·37 (10·0)	0·009 (0·25)	0·006 (0·17)	0·05 (5)
Granite	UK	1·04 (28)	0·089 (2·4)	0·085 (2·3)	0·28 (28)
	USSR	1·48 (40)	0·111 (3·0)	0·167 (4·5)	0·46 (46)
Rock aggregate	Sweden	0·81 (22)	0·048 (1·3)	0·070 (1·9)	0·20 (20)
	UK	0·81 (22)	0·052 (1·4)	0·004 (0·1)	0·12 (12)
Phospho-gypsum	UK	0·07 (2)	0·777 (21·0)	0·019 (0·5)	0·68 (68)
	USA	—	1·480 (40·0)	0·007 (0·2)	1·26 (126)
Wood	Sweden	—	—	—	<0·004 (<0·4)

Data from the Report to the United Nations Scientific Committee on the Effects of Atomic Radiation, 1977, General Assembly document 32 amount of radioactivity is wood and a wooden house does provide some degree of shielding from gamma radiation without itself contributing significantly to the dose. The absorbed dose rate on the ground floor of a wooden house is about 75% of that outdoors, and one floor up it is even lower because of the increased distance from the ground itself. An average value of 70% is therefore usually assumed for the indoor dose rate of a wooden house relative to that outside. For a house made of more solid building materials, the absorbed dose rate will be higher than that measured outdoors. The reason for this is that the building materials contain natural radionuclides and the occupant receives radiations not only from below but from all four sides and on top as well – a condition defined as 4π-geometry. A brief list of building materials is given in table 2.6. The average increase over the outdoors absorbed dose rate is taken to be 30%, but this will obviously depend very much upon the local geology and the origin of the building materials. For example, in Scotland a person living in Edinburgh would be expected to receive an absorbed dose rate in air of about 600 μGy y^{-1} (60 mrad y^{-1}) indoors, and only about 480 μGy y^{-1} (48 mrad y^{-1}) outdoors. Should he move from Edinburgh, which is built in an area of sedimentary rock, to the granite city of Aberdeen, his expected absorbed dose rate in air indoors would rise to about 850 μGy y^{-1} (85 mrad y^{-1}), but his expected absorbed dose rate in air outdoors would more than double, to about 1 mGy y^{-1} (100 mrad y^{-1}). In fact the *total* dose received varies not only because of differences in geometry but because of the inhalation of gaseous radionuclides emanating from the materials, as we shall see. It is thus obviously extremely difficult to generalize, at a population level, on the average indoor absorbed dose rate. Nevertheless estimates have been made and, making due allowance for differing construction materials, and their availability in various parts of the world, it has been estimated that the majority of the world population receives an indoor absorbed dose rate in air of 0·02 to 0·09 μGy h^{-1} (2 to 9 μrad h^{-1}).

External dose rates to man, both from exposure outdoors and indoors, therefore vary enormously. In discussing such external absorbed dose rates it would appear that no attempt is made to differentiate between the doses absorbed by different parts of the body. For internal exposure the fate of a radionuclide is particularly important; the metabolism of radionuclides generally by man will be discussed in chapter 3. It should be mentioned in passing, however, that one particular calculation of value can be made from the estimates of external dose rates and that is the annual dose absorbed by the gonads. The average annual absorbed dose for the gonads has been estimated to be

320 μGy (32 mrad), with a range for the majority of the human race of 210 to 430 μGy (21 to 43 mrad).

Radionuclides formed by the cosmic neutron bombardment of the atmosphere, and those naturally occurring, can be absorbed by man through ingestion and inhalation, resulting in *internal irradiation*. For internally accumulated radionuclides, both particle and electromagnetic radiations have to be considered, the former being the more important. In assessing the absorbed dose rates resulting from the presence of a radionuclide in a particular organ it is thus necessary to consider, and allow for, the nature and energy of the total radiation spectrum of that radionuclide.

The major naturally occurring source of internal radiation to man is ^{40}K. Potassium is the principal intracellular ion and therfore under close homeostatic control. It is not, however, uniformly distributed throughout the body. Some tissues, such as muscle, brain and blood cells contain more than 0·3% by weight of potassium whereas blood serum has a normal level of about 0·01% and fat contains virtually no potassium at all. An average value for potassium content is therefore not easily computed because much depends on the type of build of the individual. The average value used for various calculations is that an adult male contains about 2 g potassium kg^{-1} body weight – a total of 140 g for a 70 kg man. The abundance ratio of ^{40}K, as we have noted is 0·012% and therefore the 140 g potassium in man contains ~3·7 kBq (0·1 μCi) of ^{40}K. The concentrations of potassium and ^{40}K in a number of tissues and organs of particular interest are given in table

TABLE 2.7.   *Tissue concentrations (wet weight) of ^{40}K in man and the annual absorbed dose resulting from such concentrations.*

Organ or tissue	Potassium, g kg^{-1}	^{40}K Bq kg^{-1}	(pCi kg^{-1})	Annual absorbed dose μGy	(mrad)
Lung	2·1	62·9	(1700)	170	(17)
Testis	2·1	62·9	(1700)	170	(17)
Ovaries	1·35	40·7	(1100)	120	(12)
Red marrow	4·4	133·2	(3600)	270	(27)
Yellow marrow	0·6	18·5	(500)	60	(6)
Bone lining cells	—	—	—	150	(15)
Average whole-body	2	59·2	(1600)	170	(17)

Data from the Report to the United Nations Scientific Committee on the Effects of Atomic Radiation, 1977, General Assembly document 32 Session, Supplement No. 40 (A/32/40) (New York: United Nations).

2.7, together with estimates of the annual absorbed dose resulting from the radionuclide's presence. In order to calculate the latter, certain assumptions have been made on the masses of these organs in an adult male – a topic to which we will return in chapter 3 – and on a number of dosimetric factors relating to the distribution of ^{40}K in the body.

A second naturally occurring radionuclide is ^{87}Rb. Little is known of the behaviour of rubidium in the environment. The ^{87}Rb isotope has a much higher abundance ratio (27·9%) than that of ^{40}K and thus although rubidium occurs at much lower concentrations than potassium in the body it nevertheless provides a significant absorbed dose to man, as can be seen in table 2.8.

TABLE 2.8. *Tissue concentrations (wet weight) of ^{87}Rb in man and the annual absorbed dose resulting from such concentrations.*

Organ or tissue	Rubidium,	^{87}Rb		Annual absorbed dose	
	mg kg^{-1}	Bq kg^{-1}	(pCi kg^{-1})	μGy	(mrad)
Lung	9·2	8·1	(220)	4	(0·4)
Testis	20	17·8	(480)	8	(0·8)
Red marrow	7·8	7·0	(190)	4	(0·4)
Yellow marrow	7·8	7·0	(190)	3	(0·3)
Bone lining cells	—	—	—	9	(0·9)
Average whole-body	9·7	8·5	(230)	4	(0·4)

Data from the Report to the United Nations Scientific Committee on the Effects of Atomic Radiation, 1977, General Assembly document 32 Session, Supplement No. 40 (A/32/40) (New York: United Nations).

Before the rather complex uranium and thorium series are examined, brief mention may be made of the principal radionuclides formed in the atmosphere, ^{3}H, ^{7}Be and ^{14}C. The annual dose to the organs listed in tables 2.7 and 2.8 from tritium, almost entirely as tritiated water, is about 0·01 μGy (1 μrad). The majority of the ^{7}Be inventory is in the atmosphere and this contributes some 0·02 μGy (2 μrad) annually to the human lung. Naturally produced ^{14}C, however, is accumulated by all organs of the human body giving a whole-body average annual dose of about 13 μGy (1·3 mrad). This value is slightly misleading, however, in that different tissues receive different doses; for example, the annual absorbed dose for the gonad has been estimated at 5 μGy (0·5 mrad) while that for the red bone marrow has been estimated as high as 22 μGy (2·2 mrad).

The radionuclides of the uranium and thorium series contribute to the internal radiation background of man in a number of different ways. Not only are they present in his body and his food but, because some of the radionuclides in the series are gases, they can also be inhaled, such inhalation being dependent upon surrounding materials and the degree of enclosure of the environment. It is easier to discuss these radionuclides individually without having to describe continually their relevant positions in the different series; reference should therefore be made to tables 2.2, 2.3 and 2.4. Uranium, containing both ^{238}U, ^{234}U and the immediate decay products of the former (^{234}Th and ^{234m}Pa) in radioactive equilibrium, occurs in dust particles in the air as a result of being a natural constituent of soil. Both uranium isotopes emit alpha particles, whereas ^{234}Th and ^{234m}Pa are beta and gamma emitters. Typically 1 g uranium contains ~12·2 kBq (0·33 μ Ci) of each of the four nuclides. The average concentration of ^{238}U in air is taken to be ~2·6 μBq m^{-3} (~70 aCi m^{-3}) resulting in a daily adult inhalation rate of ~52 μBq (~1·4 fCi). The dietary intake of ^{238}U varies but is generally considered to be of the order of 15 mBq d^{-1} (0·4 pCi d^{-1}). This rate of intake results in soft tissue concentrations of ~7·4 mBq kg^{-1} wet (~0·2 pCi kg^{-1} wet) and bone concentrations of ~148 mBq kg^{-1} dry (~4 pCi kg^{-1} dry).

Thorium is also present in airborne dust resulting in a daily intake of ~37 μBq d^{-1} (1 fCi d^{-1}) of ^{232}Th. Dietary items contain very low concentrations of thorium and the relative importance of this route of uptake has not been fully evaluated. Bone contains ~19 mBq kg^{-1} dry (~0·5 pCi kg^{-1} dry) of ^{232}Th, and soft tissues (wet) are approximately an order of magnitude less. There are no data on ^{230}Th in man. This radionuclide has a similar concentration in soil to ^{232}Th and also has a relatively long half-life; it is assumed, therefore, that its concentration in man is similar to that of ^{232}Th. Both thorium nuclides are alpha emitters and the annual absorbed doses to a number of organs as a result of internally deposited uranium and thorium are given in table 2.9. It is worth mentioning that these doses refer to relatively clean environments. In addition to variations in natural levels of uranium and thorium in the environment, the use of some chemicals, particularly phosphate products, can enhance such levels in confined areas. Phosphate deposits contain relatively high concentrations of ^{238}U, and higher-than-average concentrations of ^{232}Th. Spread relatively thinly over agricultural land such enhanced levels result in negligible increased dose rates, but absorbed dose rates in air of up to 0·5 μGy h^{-1} (50 μrad h^{-1}) have been measured in enclosed areas where phosphate fertilizers are stored. Waste gypsum from the phosphate industry is used

TABLE 2.9. *Tissue concentrations (wet weight) of* ^{238}U, ^{234}U, ^{232}Th *and* ^{230}Th *in man and the annual absorbed dose* (α) *resulting from such concentrations.*

Organ or tissue	Concentration of nuclide, mBq kg⁻¹ (pCi kg⁻¹)		Annual absorbed dose (α), μGy (mrad)	
	^{234}U or ^{238}U	^{230}Th or ^{232}Th	^{234}U + ^{238}U	^{230}Th + ^{232}Th
Lung	7 (0·2)	19 (0·5)	0·4 (0·04)	0·8 (0·08)
Gonads	7 (0·2)	2 (0·05)	0·4 (0·04)	0·08 (0·008)
Bone†	148 (4)	—	—	—
Trabecular bone	—	48 (1·3)	—	—
Red marrow	7 (0·2)	2 (0·05)	0·5 (0·05)	0·9 (0·09)
Bone lining cells	7 (0·2)	—	3·0 (0·3)	15·0 (1·5)

†Dry weight.

Data from the Report to the United Nations Scientific Committee on the Effects of Atomic Radiation, 1977, General Assembly document 32 Session, Supplement No.40 (A/32/40) (New York: United Nations.)

as a building material and, as can be seen in table 2.6, such phospho-gypsum contributes significantly to the external absorbed dose rate to man.

Radium is another naturally occurring element of interest. There are thirteen known radioisotopes of radium but only ^{226}Ra (an alpha emitter) and ^{228}Ra (a beta emitter which gives rise to ^{224}Ra, another alpha emitter) are of environmental significance to man. As for the uranium and thorium nuclides, the inhalation of dust particles results in an intake of ~37 μBq d^{-1} (~1 fCi d^{-1}) but the average dietary intake of ^{226}Ra is quite high, of the order of 37 mBq d^{-1} (1 pCi d^{-1}). The dietary intake of ^{228}Ra is thought to be similar. The average soft tissue concentration of ^{226}Ra in man is taken to be 4·8 mBq kg^{-1} wet (0·13 pCi kg^{-1} wet). Radium behaves, chemically, in a similar manner to calcium and thus an appreciable fraction of the daily intake is deposited on bone surfaces and areas of high mineral metabolism. The distribution of radium (and calcium) in bone is thus non-uniform, making assessments of radium concentrations in bone rather meaning-less; the concentrations of ^{226}Ra range from about 74 to 740 mBq kg^{-1} dry (2 to 20 pCi kg^{-1} dry) with a mean value of about 0·3 Bq kg^{-1} dry (8 pCi kg^{-1} dry). The annual dose rate resulting from ^{226}Ra and its short-lived decay products are given in table 2.10.

Perhaps the most radiologically significant naturally occurring radionuclides are ^{222}Rn and ^{220}Rn. These two nuclides of radon arise indirectly from the decay of ^{238}U and ^{232}Th, respectively. Radon is a noble gas and emanates from the ground, and from building materials, to give ^{222}Rn and ^{220}Rn concentrations in air far in excess of their precursors. The concentrations of ^{238}U and ^{232}Th in soil are about the same but the emanation rate of ^{220}Rn, in units of radioactivity, is about 100 times as great as that of ^{222}Rn because of its higher decay constant. Their rate of emanation from any soil is influenced by the condition of the soil; that is its degree of compaction, its moisture content, its temperature and so on. In addition, a higher atmospheric pressure, increased heavy rainfall and snow cover all serve to reduce the rate of emanation. There are even diurnal variations, the day maxima being twice the night minima. Again, because of its greater decay constant, all of these factors affect the rate of emanation of ^{220}Rn more than that of ^{222}Rn. Geographic factors also affect the concentra-tions which may obtain in air because their emanation rates are much greater over land than on the sea.

It is obvious from the above that it is extremely difficult to generalize on the typical concentrations of these nuclides in air. The concentrations of both radionuclides at ground level are of the same

TABLE 2.10. *Tissue concentrations (wet weight) of ^{226}Ra and its short-lived decay products in man and the annual absorbed dose resulting from such concentrations.*

Organ or tissue	Concentration of ^{226}Ra		Concentration of short-lived decay products of ^{226}Ra		Annual absorbed dose ($^{226}Ra + ^{222}Rn + ^{218}Po + ^{214}Pb + ^{214}Bi + ^{214}Po$)			
	mBq kg⁻¹	(pCi kg⁻¹)	mBq kg⁻¹	(pCi kg⁻¹)	μGy (α)	(mrad)	μGy (β, γ)	(mrad)
Lung	5	(0·13)	2	(0·043)	0·3	(0·03)	0·01	(0·001)
Gonads	5	(0·13)	2	(0·043)	0·3	(0·03)	0·01	(0·001)
Bone†	296	(8·0)	96	(2·6)	—	—	—	—
Red marrow	5	(0·13)	2	(0·043)	0·9	(0·09)	0·1	(0·01)
Bone lining cells	—	—	—	—	7·0	(0·7)	0·3	(0·03)

†Dry weight

Data from the Report to the United Nations Scientific Committee on the Effects of Atomic Radiation, 1977, General Assembly document 32 Session, Supplement No.40 (A/32/40) (New York: United Nations.)

order of magnitude; about 3.7 mBq l^{-1} (0.1 pCi l^{-1}) in continental air, 0.37 mBq l^{-1} (0.01 pCi l^{-1}) in coastal areas and small islands, and only 37 μBq l^{-1} (1 fCi l^{-1}) over the oceans. At higher altitudes the concentration of both radionuclides decreases, ^{220}Rn more than that of ^{222}Rn.

Because of the presence of ^{238}U and ^{232}Th in building materials the concentration of radon inside buildings is considerably greater than that outdoors. The concentration will obviously depend upon the nature of the building materials and the ventilation conditions of any given room. The latter is particularly important. For example, in a room of about 30 m^3 volume a ^{222}Rn concentration of 740 mBq l^{-1} (20 pCi l^{-1}) can be reduced to 5.6 mBq l^{-1} (0.15 pCi l^{-1}) by a ventilation rate of one room change of air per hour.

The radiological significance of these radon isotopes is that they are inhaled. Their decay products are produced as free atoms, usually as positive ions because the alpha particles carry electrons away from the atom during the decay process. The ions are inclined to form clusters with water, oxygen or trace gases and the clusters thus formed tend to become attached to particles in suspension. The deposition within the respiratory tract therefore depends upon the size and type of aerosol particle to which the daughter radionuclides are attached. In estimating the absorbed dose to man it is necessary to consider the total concentration of radioacivity of the two radon isotopes, and of their daughters, at equilibrium. Using such estimated concentrations the data in table 2.11 have been derived, together with estimates of the annual absorbed dose resulting from their inhalation. The expected typical ranges are about an order of magnitude on either side of these values and thus one can see how varied such absorbed doses, particularly to the lung, may be from one individual to another. It should also be noted that ^{222}Rn in natural gas may result in an annual dose in the bronchial epithelium of 10 to 20 μGy (1 to 2 mrad).

It is of further interest to consider the long-lived decay products of ^{222}Rn, the ^{210}Pb – ^{210}Bi – ^{210}Po chain. The daily intake from inhalation of ^{210}Pb is estimated as 11 mBq d^{-1} (0.3 pCi d^{-1}) and 2.6 mBq d^{-1} (0.07 pCi d^{-1}) of ^{210}Po. For cigarette smokers, at 20 cigarettes a day, these values are increased by factors of 4 and 20, respectively. The concentrations of ^{210}Pb and ^{210}Po in food varies considerably being notably higher, 3.7 Bq kg^{-1} wet (100 pCi kg^{-1} wet) for ^{210}Po, in aquatic food species. The two elements are differently distributed in the body, lead being a bone-seeker but polonium, atypical of alpha emitters, concentrates in soft tissues.

The estimated annual absorbed doses to man at 'normal' concen-

TABLE 2.11. *Estimation of annual absorbed dose resulting from the inhalation of ^{222}Rn and ^{220}Rn short-lived decay products.*

Source of irradiation	Equilibrium equivalent concentration mBq l⁻¹(pCi l⁻¹) in air		Annual absorbed dose, μGy (mrad)							
			Lung		Gonads		Bone marrow		Bone lining cells	
^{222}Rn short-lived decay products										
Outdoor exposure	2	(0·06)	10	(1)	0·07	(0·007)	0·08	(0·008)	0·08	(0·008)
Indoor exposure from building materials	19	(0·5)	300	(30)	2	(0·02)	3	(0·3)	3	(0·3)
^{220}Rn short-lived decay products										
Outdoor exposure	0·037	(0·001)	1	(0·1)	0·002	(0·0002)	0·03	(0·003)	0·03	(0·003)
Indoor exposure from building materials	0·37	(0·01)	40	(4)	0·08	(0·008)	1	(0·1)	1	(0·1)

Data from the Report to the United Nations Scientific Committee on the Effects of Atomic Radiation, 1977, General Assembly document 32 Session, Supplement No.40 (A/32/40) (New York: United Nations.)

TABLE 2.12. *Summary of estimated annual absorbed dose, from natural sources, to man living in areas of 'normal' radiation background.*

Source of irradiation		Annual absorbed dose, μGy(mrad)			
		Gonads	Lung	Bone lining cells	Red bone marrow
External					
Cosmic rays:					
Ionizing component		280 (28)	280 (28)	280 (28)	280 (28)
Neutron component		3·5 (0·35)	3·5 (0·35)	3·5 (0·35)	3·5 (0·35)
Terrestrial radiation (γ)		320 (32)	320 (32)	320 (32)	320 (32)
Internal					
^{3}H	(β)	0·01 (0·001)	0·01 (0·001)	0·01 (0·001)	0·01 (0·001)
^{7}Be	(γ)	— (—)	0·02 (0·002)	— (—)	— (—)
^{14}C	(β)	5·0 (0·5)	6·0 (0·6)	20 (2·0)	22 (2·2)
^{40}K	($\beta + \gamma$)	150 (15)	170 (17)	150 (15)	270 (27)
^{87}Rb	(β)	8·0 (0·8)	4·0 (0·4)	9·0 (0·9)	4·0 (0·4)
^{238}U–^{234}U	(α)	0·4 (0·04)	0·4 (0·04)	3·0 (0·3)	0·7 (0·07)
^{230}Th	(α)	0·04 (0·004)	0·4 (0·04)	8·0 (0·8)	0·5 (0·05)
^{226}Ra–^{214}Po	(α)	0·3 (0·03)	0·3 (0·03)	7·0 (0·7)	1·0 (0·1)
^{210}Pb–^{210}Po	($\alpha + \beta$)	6·0 (0·6)	3·0 (0·3)	34 (3·4)	9·0 (0·9)
^{222}Rn–^{214}Po	(α)inhalation	2·0 (0·2)	300 (30)	3·0 (0·3)	3·0 (0·3)
^{232}Th	(α)	0·04 (0·004)	0·4 (0·04)	7·0 (0·7)	0·4 (0·04)
^{228}Ra–^{208}Tl	(α)	0·6 (0·06)	0·6 (0·06)	11 (1·1)	2·0 (0·2)
^{220}Rn–^{208}Tl	(α)inhalation	0·08 (0·008)	40 (4)	1·0 (0·1)	1·0 (0·1)
TOTAL		780 (78)	1100 (110)	860 (86)	920 (92)

Data from the Report to the United Nations Scientific Committee on the Effects of Atomic Radiation, 1977, General Assembly document 32 Session, Supplement No.40 (A/32/40) (New York: United Nations.)

trations of all of these radionuclides are given in table 2.12., together with a summary of absorbed doses from all other natural sources of radioactivity in the environment. It can be seen that the principal source of absorbed dose to all of the organs is external irradiation. Of the internally deposited radionuclides ^{40}K is the major source for the organs listed, with the exception of the lung for which inhaled ^{222}Rn and its short-lived daughters are important. The total annual absorbed dose to each of the organs listed may be rounded to about 1 mGy (100 mrad) per year.

Before leaving the subject of natural radioactivity, brief mention must be made of the natural radioactivity of sea water, which at 35°/$_{\circ\circ}$ contains about 0·4 g K l^{-1}, resulting in a radioactive content of 12·2 Bq (330 pCi) ^{40}K l^{-1}. This accounts for over 90% of the total radioactivity in sea water. Rubidium is also present and at 120 μg l^{-1} results in 110 mBq (3 pCi) ^{87}Rb l^{-1}. A third element of significance is uranium which, at 3·3 μg l^{-1} results in a combined uranium isotope radioactivity of 80 mBq l^{-1} (2·2 pCi l^{-1}); the concentration of uranium in sea water is sufficiently high, in fact, for its extraction as a source of nuclear fuel to have been given serious consideration. In addition to these nuclides there are the families of daughter nuclides of the uranium, thorium and actinium series, nuclides produced as a result of cosmic radiation and nuclides released as a result of the testing of nuclear weapons.

2.4 *Radioactive fallout*

The atmospheric testing of nuclear weapons – a now much diminished practice – has created a legacy of world-wide exposure to ionizing radiations for some considerable time. This is not, of course, a result of exposure from the radiations emitted at the time of detonation, but from the radionuclides which are formed and variously distributed throughout the world. The first nuclear weapon to be exploded utilized the fission of an isotope of uranium, ^{235}U. In the fission of a single nucleus of ^{235}U the neutrons released may strike other nuclei causing more fissions and the release of more neutrons. If there are sufficient ^{235}U nuclei in close proximity a chain reaction results. In a 'pure' fission bomb the fissionable material, ^{235}U or ^{239}Pu, is brought into the necessary close proximity either by bringing two or more sub-critical masses violently together or by imploding a hollow, sub-critical, mass. The result of either of these processes is a condition of supercriticality in which the neutron multiplication lasts about 10^{-8} s. In the thermonuclear, or fusion, bomb the fusion of light nuclei such as ^{2}H (deuterium) or ^{3}H (tritium) is used, creating very high temperatures.

This process also results in the creation of fast neutrons which are used to increase the explosive yield: the fusion weapon is surrounded with ^{238}U which undergoes fission as a result of the interaction with the fast neutrons.

There are two sources of radionuclides resulting from these processes. When a nucleus of ^{235}U captures a neutron it becomes a compound nucleus of ^{236}U. The nucleus then immediately disintegrates in such a way that typically about 40% of the nucleus forms one daughter nucleus and about 60% of the nucleus forms another daughter nucleus. For example, two such daughters which may be formed are one consisting of 38 protons and 57 neutrons, ^{95}Sr, and a second consisting of 54 protons and 85 neutrons, ^{139}Xe. These daughter nuclides are called *fission fragments*. A number of fission fragment pairs may be formed and they almost invariably have too many neutrons for stability. Thus each fragment starts a chain of isotopic decay. There are, in fact, about 90 possible fission fragments and their decay gives rise to over 200 radionuclides. The radionuclides arising from fission, both the primary fission fragments and the radionuclides resulting from their decay, are collectively termed *fission products*. In the above example the two fission fragments decay through two short series as follows:

$$^{95}Sr \rightarrow {}^{95}Y \rightarrow {}^{95}Zr \rightarrow {}^{95}Nb \rightarrow {}^{95}Mo$$
$$^{139}Xe \rightarrow {}^{139}Cs \rightarrow {}^{139}Ba \rightarrow {}^{139}La$$

The two stable nuclides formed, ^{95}Mo and ^{139}La, have collectively 135 neutrons, 7 neutrons having been lost via beta decay (see chapter 1) leaving a deficit of 2 neutrons. A more detailed budget analysis reveals that, in fact, some mass has been converted into energy; mass equivalent to ~0·2 atomic mass units. One atomic mass unit (m_u) is $1·6605 \times 10^{-27}$ kg, and is approximately the mass of one neutron or proton. The energy equivalent to 0·2 m_u is ~200 MeV. This energy – part of the binding energy of the large uranium nucleus – is converted principally into kinetic energy of the fission fragments. Some of it is also converted into the energy of the fission neutrons – some of which will cause the fission of other uranium nuclei – and the energy of beta particles, gamma rays and neutrinos. Occasionally – in about 0·25% of thermal-neutron fissions – three nuclei are formed, a process termed *ternary fission*. The third nucleus is usually an alpha particle but in a few cases it is tritium, ^3H. Even rarer, in about 0·001% of ^{235}U fissions, three of four nuclei of approximately similar mass are formed.

A typical fission product yield resulting from the fission of ^{235}U is

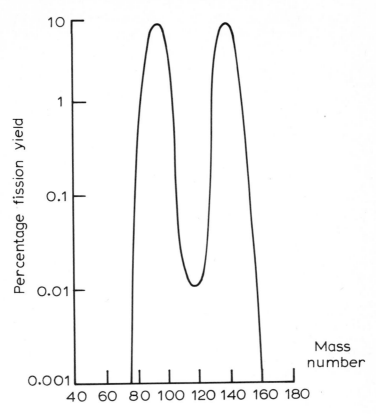

Figure 2.1.   Fission yield from ^{235}U.

shown in figure 2.1. The range of mass numbers is from about 70 to 160. The yield from the fission of ^{239}Pu shows an essentially similar pattern but is displaced two mass numbers to the right. The approximate yield of a one megaton fission bomb is given in table 2.13.

The neutron flux results not only in fissions of other uranium nuclei; neutrons may also be captured by the materials of the bomb casing and by debris sucked into the fireball which results from the explosion. The isotopes of many elements may thereby be made radioactive by neutron capture: these are the *neutron activation products*. Typical of such radionuclides are those of elements such as iron, zinc, manganese and cobalt – all of which are essential trace elements. The type, and amounts, of neutron activation products formed will depend upon where, and ·at what height, the bomb was exploded. Elements in the atmosphere are also subject to neutron interactions,

TABLE 2.13. *Approximate yields of some of the principal radionuclides resulting from the fission of a megaton nuclear weapon.*

Nuclide	Half-life	PBq (MCi)			
		at 1 month		at 1 year	
^{89}Sr	50·5 d	407	(11)	4·8	(0·13)
^{90}Sr	28·5 y	3·7	(0·1)	3·7	(0·10)
^{95}Sr	64 d	555	(15)	16	(0·43)
^{131}I	8 d	233	(6·3)	0	(0)
^{137}Cs	30 y	5·9	(0·16)	5·9	(0·16)
^{144}Ce	284 d	130	(3·5)	59	(1·6)
^{45}Ca	164 d	9·6	(0·26)	2·2	(0·06)
^{55}Fe	2·7 y	3·7	(0·10)	3·3	(0·09)
^{14}C	5730 y	1·1	(0·03)	1·1	(0·03)

y = years, d = days.

From Klement, A.W., Jr, (1965), *Health Physics* 11, 1265 reproduced by permission of the Health Physics Society.

notably nitrogen, which produces some 1·26 PBq (34 kCi) of ^{14}C per megaton. A much smaller amount of ^{3}H is also produced.

Debris which enters the fireball soon after its formation is vaporized. Subsequent cooling, resulting from the expansion and rising of the fireball, causes the volatilized debris to condense, forming an aerosol with a wide distribution of particle sizes. Oxides of iron, aluminium and other refractory materials form particles of 0·4 to 4·0 μm diameter; whereas more volatile radionuclides condense into smaller particles of less than 0·4 μm diameter. This fractionation of formed particles results in a differential deposition of the particles near the explosion site. Thus radionuclides present in the larger aerosol particles are deposited within a few hundred kilometres and constitute the local fallout. Smaller particles may be injected into the troposphere and be transported, mainly latitudinally, around the Earth; these particles constitute the tropospheric fallout. Finally, those particles which are carried up into the stratosphere constitute stratospheric, or global, fallout; although again the major deposition, as in tropospheric fallout, will take place in the same hemisphere as the site of the explosion. The relative amounts injected into either the troposphere or the stratosphere depend both upon the yield of the explosion and the latitude at which it occurred. Some mixing within the stratosphere and troposphere does take place. Stratospheric mixing processes are usually slower than those of the troposphere, and are horizontal.

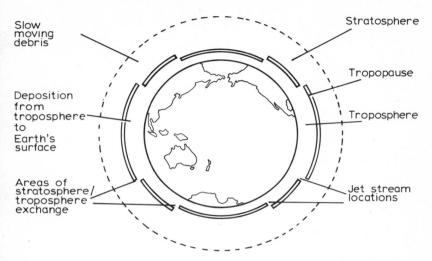

Figure 2.2. The Earth's atmospheric regions and their relationship to atmospheric fallout. Some exchange between the stratosphere and the troposphere takes place at all gaps in the tropopause, but the greatest exchange takes place at those gaps which are nearest the equator in each hemisphere.

Some degree of vertical mixing between the two atmospheric layers also occurs – particularly via the sub-tropical tropopause gap (figure 2.2). There is a seasonal variation too, with a maximum occurrence in late winter, so that fallout deposition is usually highest in the spring. Radionuclides injected into the stratosphere stay there for variable periods of time. For example, if the nuclides were injected into the stratosphere at an altitude of ~20 km, the average residence time for a radionuclide such as ^{90}Sr would be about 1 year, although it could vary from about 4 months to 2 years depending upon the latitude and time of year of the explosion. Some radionuclides, such as ^{14}C, appear to move both from the stratosphere to the troposphere and in the opposite direction, processes which considerably prolong their ultimate deposition.

Radionuclides in the troposphere, either from direct injection or from the stratosphere, are dispersed relatively quickly thoughout the hemisphere. Gaseous radionuclides, such as ^{85}Kr, exchange between the northern and southern hemispheres; but particulate radionuclides are deposited within the hemisphere in which the explosion took place. The radionuclides are deposited by three different processes: the particles may take part in the formation of droplets within clouds, the falling rain drops may collect particles as they fall or, of particular importance in arid regions, the particulate matter may settle out by dry deposition. As a result of these processes particulate radionuclides only remain in the troposphere for an average residence time of about 1 month.

Local fallout, at the site of detonation, produces intense radiation. The quantity of radioactivity in the debris produced in a nuclear explosion decreases to one twentieth during the period of 1 to 24 h after detonation. A further time-lag of 1 month in the troposphere considerably reduces the quantity of tropospheric fallout. One radionuclide of particular importance here is ^{131}I which, although it has a half-life of only 8 days, can cause considerable exposure to human beings, particularly children. This is because fresh milk dominates as a source of ^{131}I, and ^{131}I concentrates in the thyroid gland. Small children not only have a high intake of milk but have relatively larger thyroid glands than adults.

The lasting, world-wide, effects of stratospheric fallout result from the presence of a relatively small number of radionuclides. Of these the two most important have been ^{137}Cs, with a half-life of 30·1 years, and ^{90}Sr, with a half-life of 28·5 years. Both of these are fission products. Neutron activation products of importance are ^{239}Pu – which has a half-life of $2·4 \times 10^4$ years and is produced by neutron capture by ^{238}U – and ^{14}C and ^3H.

Fallout radionuclides give rise to both external and internal irradiation of man. The major long-term source of external radiation is ^{137}Cs. The dose rates resulting from both the gamma and beta components of fallout have been estimated for a number of areas on the Earth's surface. Annual mean values for one location, Chilton, Oxfordshire, England, are given in table 2.14 for the years 1951 to 1977. The values are estimated at a height of 1 m above the ground and are for air-absorbed doses. The relatively high values resulting from the large number of atmospheric tests in the late 1950s and immediately before the 1963 partial test ban treaty can clearly be seen. The natural levels at 1 m above the ground in this area are estimated as ~0·616 mGy (61·6 mrad) per annum of gamma radiation and ~0·122 mGy (12·2 mrad) per annum of beta radiation. The greatest relative increase is that of beta radiation. The importance of short-lived radionuclides such as ^{95}Zr, ^{103}Ru, ^{106}Ru, ^{141}Ce, ^{140}Ba, ^{144}Ce to external radiation is also apparent.

Internal radiation to man results from the accumulation of both fission and neutron activation products. The effects of such accumulation depend upon a large number of factors which include the chemical nature of the radionuclide, the type of radiation emitted,

TABLE 2.14. *Absorbed annual dose rate in air, 1 m above the ground, at Chilton, Oxfordshire, England since 1951 due to fallout.*

Year	Rainfall, mm	Total dose at 1 m, μGy (mrad)			
		Gamma		Beta	
1951	823	46	(4·6)	240	(24·0)
1952	680	35	(3·5)	225	(22·5)
1953	535	76	(7·6)	570	(57·0)
1954	712	40	(4·0)	462	(46·2)
1955	511	43	(4·3)	715	(71·5)
1956	583	62	(6·2)	878	(87·8)
1957	690	113	(11·3)	1299	(129·9)
1958	804	126	(12·6)	1584	(158·4)
1959	703	172	(17·2)	1881	(188·1)
1960	969	36	(3·6)	710	(71·0)
1961	639	79	(7·9)	632	(63·2)
1962	629	230	(23·0)	1635	(163·5)
1963	643	251	(25·1)	3229	(322·9)
1964	473	89	(8·9)	2227	(222·7)
1965	640	73	(7·3)	1142	(114·2)
1966	817	50	(5·0)	434	(43·4)
1967	752	36	(3·6)	209	(20·9)
1968	716	30	(3·0)	177	(17·7)
1969	557	30	(3·0)	192	(19·2)
1970	725	29	(2·9)	224	(22·4)
1971	695	28	(2·8)	244	(24·4)
1972	616	25	(2·5)	258	(25·8)
1973	552	24	(2·4)	102	(10·2)
1974	800	22	(2·2)	78	(7·8)
1975	568	21	(2·1)	83	(8·3)
1976	521	24	(2·4)	91	(9·1)
1977	819	25	(2·5)	127	(12·7)
Natural levels		616	(61·6)	122	(12·2)

The author thanks the United Kingdom Atomic Energy Authority (UKAEA) for permission to quote from Table 25 of *Radioactive Fallout in Air and Rain — Results to the end of 1977 (AERE-R 9016)* by R.S. Cambray *et al.*

TABLE 2.15. *Estimates of the average, integrated, deposition density - without taking account of radioactive decay - of fallout ^{90}Sr, to the end of 1975.*

Latitude band	Area of band (10^6 km^2)	Relative population of band (%)	Integrated deposition density of ^{90}Sr,	
			MBq km^{-2}	(mCi km^{-2})
70° − 80°N	11·6	0·0	707	(19·1)
60° − 70°N	18·9	0·4	1691	(45·7)
50° − 60°N	25·6	12·2	2930	(79·2)
40° − 50°N	31·5	13·8	3149	(85·1)
30° − 40°N	36·4	18·2	2309	(62·4)
20° − 30°N	40·2	29·2	1691	(45·7)
10° − 20°N	42·8	9·8	1110	(30·0)
0° − 10°N	44·1	5·6	766	(20·7)
0° − 10°S	44·1	5·8	699	(18·9)
10° − 20°S	42·8	1·8	374	(10·1)
20° − 30°S	40·2	1·6	666	(18·0)
30° − 40°S	36·4	1·4	766	(20·7)
40° − 50°S	31·5	0·1	907	(24·5)
50° − 60°S	25·6	0·05	544	(14·7)

Data from the Report to the United Nations Scientific Committee on the Effects of Atomic Radiation, 1977, General Assembly document 32 Session, Supplement No.40 (A/32/40) (New York: United Nations.)

the half-life of the nuclide, its ultimate location in the body, and the rate at which the body excretes it. Many of these important factors will be discussed in chapter 3 and it will suffice here to note that some of the more important radionuclides resulting from weapons testing, like naturally occurring radionuclides, will give rise to different absorbed doses to different parts of the body. One of the more important radionuclides, universally associated with fallout, is ^{90}Sr. This radionuclide also provides one of the best examples of latitudinal differences in fallout deposition, as can be seen from table 2.15. Such data are of considerable value in estimating the dose to different populations, and to the total world population. They also serve as a guide to the latitudinal distribution of other radionuclides occurring in fallout.

At any given latitude the amount of ^{90}Sr deposited on to the Earth's surface is affected by the amount of rainfall. The average, estimated, deposition of ^{90}Sr over the United Kingdom since 1958 is shown in figure 2.3., together with the cumulative deposition. The ^{90}Sr is taken up by plants, both from the soil and as a result of direct foliar contamination. The dietary intake of ^{90}Sr is therefore dependent upon food habits, but milk is an almost universally important source and has been a good index of exposure to children in different countries.

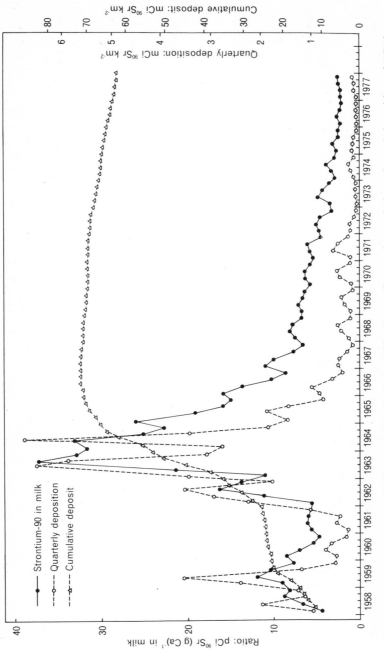

Figure 2.3. Average ratio of ^{90}Sr to calcium in United Kingdom milk samples, and the estimated deposition of ^{90}Sr in fallout. (1 pCi = 37 mBq) From the 1977 Annual Report of the Agricultural Research Council's Letcombe Laboratory; courtesy of Dr R.S. Bruce and of the director of the laboratory.

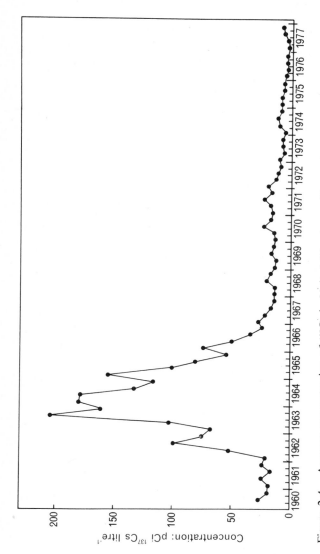

Figure 2.4. Average concentration of ^{137}Cs in United Kingdom milk samples. (1 pCi = 37 mBq). An updated version of a figure from the 1976 Annual Report of the Agricultural Research Council's Letcombe Laboratory; courtesy of Dr R.S. Bruce and of the director of the laboratory.

Strontium behaves in a similar chemical manner to calcium and therefore the amount of ^{90}Sr in a given food item relative to its calcium content provides a very useful comparative unit, the *strontium unit* or SU. The amount of ^{90}Sr per g calcium in milk since 1958 is also shown in figure 2.3. Apart from milk the main dietary sources of ^{90}Sr are milk products, grain products, fruit and vegetables. Concentrations of ^{90}Sr in the vertebrae of adults reached maximum values in the northern hemisphere of about 0·11 Bq (g Ca)$^{-1}$ (3 pCi (g Ca)$^{-1}$) during the period of 1964 to 1966.

Another isotope of strontium, ^{89}Sr, has a much shorter half-life – 50·5 days as opposed to 28·5 years for ^{90}Sr. As a consequence it is of greater relative importance in the first few months after an atmospheric nuclear explosion, and is also more relatively important as a direct contaminant of foliage.

In addition to being a source of external irradiation, ^{137}Cs is also an important source of internal irradiation. Caesium, behaving in a chemical manner similar to that of potassium, is taken up by plants and thus occurs in a variety of foods, particularly grain and meat products. Like ^{90}Sr, ^{137}Cs has a fairly long half-life (30·1 years) and the ^{137}Cs/^{90}Sr yield quotient of 1·6, on an activity basis, has remained fairly constant in estimates of global deposition. Because of the concern over radionuclide accumulation by children, the ^{137}Cs content of milk has been carefully monitored in various parts of the world. The average values obtained in the United Kingdom since 1960 are shown in figure 2.4. Concentrations of ^{137}Cs are frequently expressed in units relative to potassium in a manner analogous to that of the strontium unit.

The radionuclides of strontium and caesium are of importance because these elements are chemically similar to elements which are essential to living organisms. Two radionuclides *of* essential elements are ^{14}C and ^{3}H, both of which are therefore readily accumulated. The ^{14}C in the atmosphere is present as carbon dioxide and thus taken up by plants during photosynthesis and eventually, via various food chains, incorporated into human tissues. The majority of ^{3}H is in the form of tritiated water which also becomes incorporated, with a fairly even distribution, in human tissues.

There are a number of other radionuclides which may be mentioned. Two fairly short-lived ones, ^{106}Ru and ^{144}Ce, are not readily incorporated into tissues but may be inhaled and give rise to irradiation of the lung. The long-lived radionuclide ^{239}Pu is also inhaled, not only irradiating the lung but being absorbed into the body by this route and deposited in bone. The radionuclides of metals, such as ^{54}Mn and ^{55}Fe, may also be inhaled.

In assessing the dose to man as a result of all of these radionuclides in fallout, it is necessary to adopt a different approach from that used for natural radionuclides. Natural radionuclides, although varying from place to place, are present all the time. One can therefore calculate an annual absorbed dose. But, as we have seen, the concentrations of radionuclides in fallout vary not only spatially but also vary with time. Thus in order to assess the absorbed dose to man in a manner with which comparisons of absorbed dose from any other source could be compared, it is necessary not only to assess the dose which a population receives now, but the dose that it is committed to receive in the future. It is worth digressing for a moment to see how such an assessment can be made.

In any exposed population there will obviously be a range of individual dose rates. It is difficult to represent the dose rate to the population by any single value, but an attempt is made by assessing the 'weighted' product of dose rates due to a particular source, such as fallout, and the number of individuals in the exposed population. This is the *collective dose rate* ($\dot{S}$) defined as

$$\dot{S} = \int_0^\infty \dot{D} N_{\dot{D}}(\dot{D}) \, d\dot{D} \tag{2.1}$$

where $N_{\dot{D}}(\dot{D})$ is the population spectrum in dose rate, and $N_{\dot{D}}(\dot{D}) \, d\dot{D}$ is the number of individuals receiving a dose rate in a specified organ or tissue, due to that source, in the range of $\dot{D}$ to $\dot{D} + d\dot{D}$.

Having obtained a collective dose rate one can then integrate this value over a specified period of time to obtain a *collective dose* (S). The time integral can be made for a few individuals, a population group, or the total population. The collective dose can also be assessed as ·the weighted product of individual doses and the number of individuals, in a manner similar to that of equation (2.1). The average dose ($\overline{D}$) in an exposed population of N individuals can also be derived. This dose, called the *per caput* dose, is most simply expressed as S/N for a given population exposed to a certain source of radiation. It also follows that a *per caput* dose rate can similarly be defined.

In order to give an estimate of the *total* exposure to be expected, however, it is necessary to extend the calculation in time. Two methods are used: the infinite time integral of the collective dose rate, termed the *collective dose commitment*; and the infinite time integral of the *per caput* dose rate, termed the *dose commitment*.

For internal irradiation the dose commitment is calculated as

$$D^C = K \int_0^\infty \overline{Q}(t) \, dt \tag{2.2}$$

where $\overline{Q}$ (t) is the *per caput* estimate of the quantity of radioactivity in the body or tissue at time (t) and K is a conversion factor appropriate to a particular population.†

The dose commitments for different nuclides in fallout arising from nuclear weapons tests in the atmosphere before 1976, to the world population, are given in table 2.16. The total world population's dose commitment is, in fact, heavily biased towards the northern hemisphere because not only does it have the largest fraction of the world's population but the majority of weapons tests have been made in this hemisphere. The main contributor to the dose commitment in the long term is ^{14}C, irradiation from which will extend over many thousands of years owing to its half-life of 5730 years. The dose commitment from ^{137}Cs, which contributes to both external and internal irradiation, will be almost completely delivered in several decades.

2.5. *Medical irradiation*

Unlike natural radioactivity and fallout, exposure to radiation resulting from medical practice is limited to the populations of developed countries: it does not, therefore, contribute to the dose received by the world's population in a manner equivalent to either of the above sources of exposure. But it is to be imagined that any country in a position to develop a nuclear power programme will also possess medical radiological facilities and this particular source of radiation background to its population is likely to be a very significant one. Exposure to medical radiations is also an interesting area of study because it is an example of deliberate exposure to ionizing radiations by the general public on a risk/benefit basis; although any risk involved is unlikely to be appreciated by the individual involved. The risk/benefit equation of exposure to radioactive fallout is evaluated with even less individual consent! It is to be assumed, however, that exposure to radiation for medical diagnostic purposes is willingly accepted and it is of interest to see how the absorbed doses received

† The definitions given here are those used by the United Nations Scientific Committee on the Effects of Atomic Radiation (UNSCEAR), and are for absorbed dose, for which the units are in Gy (or rad). The *collective dose* is sometimes expressed as man-Gy (or man-rad) to distinguish it from individual dose. For radiological protection purposes the International Commission on Radiological Protection (ICRP), as discussed in chapters 3 and 5, uses the term *dose-equivalent* (*H*), instead of dose, for which the units are Sv (or rem). The values are numerically the same for X-rays, gamma rays and beta particles, but will differ for neutrons, protons and alpha particles because of their different quality factors as described in chapter 1. A list of definitions, as used in radiological protection, is given in Appendix 4.

TABLE 2.16. *Summary of dose commitment μGy (mrad) from radionuclides produced in all nuclear tests made before 1976 to the world population.*

Source of radiation	Gonads		Bone marrow		Bone-lining cells		Lung	
External								
Short-lived								
nuclides	300	(30)	300	(30)	300	(30)	300	(30)
^{137}Cs	380	(38)	380	(38)	380	(38)	380	(38)
Internal								
^{3}H	20	(2)	20	(2)	20	(2)	20	(2)
^{14}C†	70	(7)	320	(32)	290	(29)	90	(9)
^{55}Fe	7	(0·7)	4	(0·4)	7	(0·7)	7	(0·7)
^{90}Sr			520	(52)	710	(71)		
^{89}Sr			3	(0·3)				
^{106}Ru							240	(24)
^{137}Cs	170	(17)	170	(17)	170	(17)	170	(17)
^{144}Ce							380	(38)
^{239}Pu					9	(0·9)	9	(0·9)
Total	950	(95)	1700	(170)	1900	(190)	1600	(160)

†For dose accumulated up to the year 2000

Data from the Report to the United Nations Scientific Committee on the Effects of Atomic Radiation, 1977, General Assembly document 32 Session, Supplement No.40 (A/32/40) (New York: United Nations.)

from, for example, medical X-ray examinations compare with the annual absorbed dose values of tables 2.12 and 2.16. The whole-body dose varies considerably, depending upon the region being examined, and apart from X-ray examinations of the limbs there is considerable exposure of the more radiosensitive organs. It should also be noted that frequently more than one X-ray exposure is made per examination because of faulty exposure, or positioning faults of the patient.

The number of people receiving X-ray and other diagnostic radiation treatment is increasing in most countries, the greatest increase being that of dental X-rays. The annual, *per caput*, doses resulting from various typical types of diagnostic X-rays in Sweden – the results of a study made in 1974 – are given in table 2.17. The total, whole-body value, which includes examinations not listed in the table, was an annual 1 mGy (0·1 rad) *per caput*. The data given in table 2.17, although instructive, do not really help in an evaluation of the significance of diagnostic irradiation at the population level. One method is to try to evaluate the significance of different types of examination to possible genetic effects on the population. Genetic effects can only be manifest through children but the frequency, and type, of medical radiation received throughout life differs markedly from one person to another. Their relative significance, therefore, has to be weighted to allow for the age – and thus the chances of having children – of the persons concerned. It is nevertheless, despite the obvious difficulties, possible to calculate a *genetically significant dose* (GSD).This is defined as the dose which, if received by every member of the population, would be *expected* to produce the same genetic injury to the population as do the *actual* doses received by the various individuals. It is calculated, as an annual dose, in the following manner. The male and female contributions to

TABLE 2.17. *Annual per caput doses to organs in various diagnostic X-ray examinations in Sweden. Values given as µGy per caput (mrad per caput).*

Examination	Whole body		Ovary		Testis		Active marrow	
Hip and femur	32	(3·2)	70	(7·0)	280	(28·0)	47	(4·7)
Pelvis	19	(1·9)	29	(2·9)	48	(4·8)	29	(2·9)
Lumbosacral region	4	(0·4)	5	(0·5)	3	(0·3)	3	(0·3)
Stomach, upper intestine	130	(13·0)	17	(1·7)	5	(0·5)	120	(12·0)
Small intestine	10	(1·0)	6	(0·6)	3	(0·3)	12	(1·2)
Colon	138	(13·8)	110	(11·0)	85	(8·5)	150	(15·0)
Abdomen	39	(3·9)	26	(2·6)	26	(2·6)	39	(3·9)
Dorsal spine	40	(4·0)	<13	(<1·3)	<3	(<0·3)	62	(6·2)
Lung, ribs	35	(3·5)	<3	(<0·3)	<3	(<0·3)	32	(3·2)
Head, sinus	42	(4·2)	0	(0·0)	0	(0·0)	53	(5·3)
Femur (lower two-thirds)	4	(0·4)	3	(0·3)	24	(2·4)	0	(0·0)
Arm	4	(0·4)	0	(0·0)	0	(0·0)	0	(0·0)
Dental (single exposure)	44	(4·4)	0	(0·0)	0	(0·0)	15	(1·5)

Data from the Report to the United Nations Scientific Committee on the Effects of Atomic Radiation, 1977, General Assembly document 32 Session, Supplement No. 40 (A/32/40) (New York: United Nations).

TABLE 2.18.   Data used in calculating the genetically significant dose (GSD) resulting from X-ray examinations in Japan, 1969.

Type of examination	Frequency $(N_f^*/N) \times 1000$		Gonad dose (d_f^*) µGy (mrad)		GSD (D_f^*) µGy (mrad)		Total GSD (D_f) µGy (mrad)	
	Male	Female	Male	Female	Male	Female	Total	%
Intestine	10	8	8000 (800)	4000 (400)	39 (3·9)	27 (2·7)	66 (6·6)	24·9
Stomach	157	96	110 (11)	1300 (130)	9·6 (0·96)	55 (5·5)	64·6 (6·46)	24·4
Hip joint	7	7	4600 (460)	1200 (120)	20 (2·0)	3·2 (0·32)	23·2 (2·32)	8·7
Lumbosacral spine	5	3	5300 (530)	1800 (180)	20 (2·0)	2·6 (0·26)	22·6 (2·26)	8·5
Lumbar spine	19	10	700 (70)	2200 (220)	8·4 (0·84)	8·8 (0·88)	17·2 (1·72)	6·5
Bladder	3	2	9900 (990)	1600 (160)	12 (1·2)	1·2 (0·12)	13·2 (1·32)	5·0
Pelvis	2	2	8300 (830)	2000 (200)	11 (1·1)	2 (0·2)	13·0 (1·30)	4·9
Chest	484	408	2 (0·2)	6 (0·6)	7·3 (0·73)	4·3 (0·43)	11·6 (1·16)	4·4
Obstetrical abdomen		2		2500 (250)		6·7 (0·67)	6·7 (0·67)	2·5
Pelvis		1		4600 (460)		2·5 (0·25)	2·5 (0·25)	0·9
Other	127	76			13·8 (1·38)	11 (1·1)	24·8 (2·48)	9·3
Total	814	615			141 (14·1)	124 (12·4)	265 (26·5)	100
	1429							

Data from the Report to the United Nations Scientific Committee on the Effects of Atomic Radiation, 1972, General Assembly document 27 Session, Supplement No. 25 (A/8725) (New York: United Nations).

the GSD, D^*, where the asterisk denotes the sex, from different types, or classes, of examinations, j, are derived from

$$D^*_j = d^*_j \cdot \frac{N^*_j}{N} \cdot \frac{w^*_j}{w} \qquad (2.3)$$

where D^*_j is the contribution from class j examination, d^*_j is the mean gonad dose per individual undergoing class j examination, N^*_j/N is the relative frequency of class j examination; that is the number of examinations *per caput* per year, and w^*_j/w is the relative child expectancy of the average individual undergoing class j examination.

As an example of how such data are compiled, the data in table 2.18 are those obtained from a survey made in Japan in 1969. The principal type of examination, that of the chest, contributed to only a small fraction of the GSD.

The use of radiopharmaceuticals in diagnostic medicine is continually increasing and this technique will also contribute to the GSD. Neither these nor X-ray techniques compare, however, with the very much higher doses of radiation given in the therapeutic uses of radiation. Radiation is used in the treatment of two major classes of disease: those of the skin and other non-neoplastic diseases, and those of neoplastic diseases, which include all forms of cancer and other invasive and malignant diseases. Localized doses of up to 10 to 20 Gy (1 to 2 krad) may be given for the former, usually in the form of low-energy, non-penetrating radiation, while for neoplastic diseases very localized doses of up to 60 to 70 Gy (6 to 7 krad) may be used. In both cases such doses are carefully delivered over a period of time, that is *fractionated*, so as to give maximum effect to the diseased area whilst minimizing damage to healthy tissues. It will be evident, nevertheless, that the absorbed dose received by surrounding tissues, particularly those between the source and the target area, will be considerable; but the seriousness of the disease usually necessitates that any deleterious late effects, although minimized, are of secondary importance.

2.6. *Consumer products*

A variety of consumer products contain radioactive materials; they include radioluminous products such as watches and compasses, electronic and electrical devices, antistatic devices, smoke and fire detectors, a variety of scientific instruments – all of which include radionuclides which have been deliberately incorporated for a specific purpose. Modern watches usually contain ^3H and it has been established that tritiated water, or tritiated organic molecules, can emanate slowly from the ^{3}H-painted surfaces and become absorbed by the body. (Tritium is a pure β^- emitter and thus presents no external source of irradiation when confined within the watch.) Watches contain typically about 185 MBq (5 mCi) of ^{3}H resulting in an annual whole-body dose, from the tissue-absorbed tritium, of about 5 μGy (0·5 mrad).

Television receivers, particularly colour television receivers, are the most common electronic products with the potential of exposing the general public to external radiation; they emit X-rays. The absorbed dose is estimated to be very low, for example on average $\sim 2 \mu$Gy y^{-1} ($\sim$0.2 mrad y^{-1}) to the male gonad, and will depend upon viewing habits – both hours watched and at what distance from the screen.

It must also be noted that exposure to controlled levels of radiation is accepted in many areas of work unconnected with the nuclear industries. The most obvious example is that of the hospital radiologist, but a list of typically exposed occupational groups would include other specialists in the medical and veterinary sciences, dentists, welders – who use autoradiographic techniques to check welds – and scientists both in industrial and research establishments. There is thus an overall requirement – both for the public in general and for particular occupational groups – for guidelines and standards relating to the acceptable safe limits of exposure to radiation, taking due consideration for the radiation background in which we all live.

3. Radiation and man

3.1. *Introduction*

With or without the development of nuclear power for the generation of electricity, the production and use of diverse sources of ionizing radiations in science and technology, and particularly in medicine, has demanded a full evaluation of the effects of radiation on man in order to develop a safe code of practice for the deployment of these sources in numerous fields. Thus it has already been accepted that radionuclides, and other sources of ionizing radiations, will be used under controlled conditions for a variety of purposes; the codes of practice which have been compiled have not, therefore, been derived solely for the purpose of evaluating and controlling the impact of a nuclear power programme, although there is no doubt that the use of nuclear energy has been a major force in the rapid development of standards. Before discussing the potential hazards resulting from nuclear power, however, it is first of all necessary to understand something of the basis upon which these potential hazards of radiation can be assessed and to appreciate that, although complex, evaluation of these hazards can generally be reduced to a reasonable basis of numerical solution.

3.2. *Effects of radiation on man*

The hazards of ionizing radiations became apparent soon after the discovery of X-rays by Röntgen in 1895 and the discovery of radioactivity by Becquerel the following year, although a number of workers had been exposed to laboratory sources of radiation for a number of years prior to these discoveries. It soon emerged that persons handling radioactive materials, or exposed to X-rays, developed dermatitis followed by skin desquamation. In a number of cases cancer also developed, necessitating whole or partial amputation of the limb. The ability of radiations to cause such severe damage to tissues was also quickly recognized as a technique for selectively destroying diseased tissues – particularly neoplastic ones, which appeared to be more sensitive than healthy tissues. Early experience of the effects of radiation also stemmed from the use of radium in luminous paints during and immediately following the first world war; workers employed to paint the dials of watches ingested radium as a result of 'pointing' the tips of the brushes with their lips. Many of the workers subsequently developed bone cancers and aplastic anaemia – the suppression of all blood-cell formation – in addition to cancers of the mouth and lips. Radium was also once used medically for the treatment of a variety of ailments. From these early, unfortunate, episodes other effects were noted. At moderate doses, either temporary or permanent sterility occurred; and at high doses a series of effects resulted which were collectively described as *acute radiation syndrome*. It was gradually appreciated that the multiplicity of all of the radiation effects observed were related to a number of factors, including whether there was an external or internal source of exposure, and whether the dose received resulted from a single exposure or from a series of repeated exposures over an extended period of time. It was further noted that the radiation effects observed were either fairly quickly made apparent or did not develop for some considerable time.

We now have much more detailed information on the effects of radiation on man and new data are continually being derived and evaluated. These data arise, in addition to the above-mentioned sources, from the survivors of the nuclear bombs dropped on the Japanese cities of Hiroshima and Nagasaki in 1945; from patients who have been treated by irradiation for the treatment of various diseases such as ankylosing spondylitis (loss of movement in a joint); from children unwittingly irradiated *in utero* as a result of pelvic X-ray examinations in pregnancy; from children once irradiated with X-rays for the treatment of thymic enlargement; from radiologists who absorbed repeated small doses throughout their working life; and from a variety of individual accidents, of which the accurate documentation and detailed follow-up have been of considerable value. The process of amassing data on persons exposed to relatively low doses of radiation as part of their normal working regime continues today, and will continue to justify, or call into question, our present radiation protection standards. Data obtained on man himself are obviously the most valuable, but much has been learned from experiments on mammals in the laboratory.

In order to provide some idea of scale to the acute-dose/effect relationship for man, the information in table 3.1. has been set out in six

TABLE 3.1. *The effects of acute exposure to radiation on man.*

	Dose received, Sv (rem)†					
	>50 (>5000)	50-10 (5000-1000)	10-6 (1000-600)	6-2 (600-200)	2-1 (200-100)	<1 (<100)
Death at:	2 days	2 weeks	2 months	2 months	—	—
Cause of death:	Respiratory failure; brain oedema	Loss of fluid and essential salts	Haemorrhage; infection	Haemorrhage; infection	—	—
Incidence of death:	100-90%	100-90%	100-80%	80-0%	0%	0%
Principal affected organs:	Central nervous system	Gastrointestinal tract	Haematopoietic tissue	Haematopoietic tissue	Haematopoietic tissue	None
Characteristic signs:	Vomiting; convulsions; tremor; ataxia;‡ lethargy	Vomiting; diarrhoea; fever	Severe leukopaenia;§ purpura;¶ haemorrhage; infection; loss of hair	Severe leukopaenia;§ purpura;¶ haemorrhage; infection; loss of hair	Moderate leukopaenia§	None

†Equivalent to Gy(rad) for γ and β irradiation.
‡Inability to co-ordinate muscular actions.
§ Decrease in number of leucocytes in the blood.
¶ Bruising of the skin.

From Eisenbud, M., 1973, *Environmental Radioactivity*, 2nd edn., (London: Academic Press); courtesy of the author and publisher.

broad bands of absorbed dose. The absorbed dose has been given in sieverts (rems in parentheses) to   allow for the relative effects of different types of radiation. For comparison with the tabulated data given in chapter 2, the values are equivalent to grays (rads in parentheses) for gamma and beta radiation. Whole-body exposure to massive doses of radiation, received instantaneously or within a relatively short time, results in immediate damage to the respiratory system and to the central nervous system. At doses in excess of 50 Sv (5000 rem) death occurs within one or two days. At somewhat lower acute doses, between 50 and 10 Sv (5000 and 1000 rem), the cells lining the gut are killed. The subsequent desquamation of the gut epithelium, and the failure of cell replacement, results in an inability to absorb food or liquids from the gut lumen. This results in vomiting, severe diarrhoea and fever, death usually occurring within two weeks – primarily through severe loss of fluids and essential elements such as sodium. Lower doses are still fatal – almost certainly so if they are above 8 Sv (800 rem) – but the primary cause here is damage to the bone marrow, the resultant physiological consequences of which are collectively termed the *haemopoietic syndrome*. The onset of this disease is usually fairly sudden; it begins with nausea and vomiting but develops into a general state of malaise and fatigue, loss of hair, bruised skin, haemorrhage and virtually no natural defence against infection. As a result of these effects death usually occurs within two months. The somewhat critical dose, at which death and survival are fairly evenly balanced, appears to be about 4 Sv (400 rem). The reasons for this are not entirely clear but are probably related to the ability of the bone marrow –whose cells may be virtually ablated – to regrow spontaneously.

The bone marrow is the site of formation of red blood cells (erythrocytes). Absorbed doses down to about 2 Sv (200 rem) usually result in some degree of damage to the blood system which may, or may not, result in death. Blood in man consists of about 55%, by volume, of plasma and about 45% erythrocytes, leucocytes (white blood cells) and platelets. The erythrocytes carry oxygen from the lungs to the tissues and carbon dioxide from the tissues to the lungs. Leucocytes, of which there are a number of different types, such as granulocytes and lymphocytes, provide the body's defence against infection. The former, like erythrocytes, are formed in the bone marrow whereas lymphocytes are produced in the lymph nodes and spleen. The blood platelets are concerned with the formation of blood clots. Each of these different types of blood cell responds differently in time to sub-lethal doses of radiation. The initial response is usually an immediate increase in the number of granulocytes and a decrease in the number of lymphocytes. Within a day, however, the number of granulocytes also decreases and the numbers of both types of leucocytes remains depressed for periods of several weeks. In contrast the response of the erythrocytes is much slower, reflecting their much longer life-time in the blood, which is of the order of several months in comparison with a life-time of a matter of days for the leucocytes. A decrease in the number of circulating erythrocytes is not observed, therefore, until about a week after exposure but thereafter the decrease continues for up to two months after which, at low doses, recovery usually ensues. The blood platelet count drops steadily to a minimum at about one month, followed by a very slow rate of recovery. The extent, and rate of change, of the responses of these components of the blood to radiations have been shown to be remarkably dose-dependent.

Below doses of 1 Sv (100 rem) there is no death in the short-term. There are, in fact, a number of treatments available to persons who have received such high doses of radiation, the forms of treatment – antibiotics, blood transfusions, bone-marrow transplants – being largely those which redress the haemopoietic syndrome.

All of these acute effects of radiation on man occur at irradiation levels which are orders of magnitude greater than those discussed in chapter 2. There are, in addition, a number of delayed effects of irradiation; some also result from acute exposure, but others become more significant when the dose is delivered in repeated small exposures over an extended period of time. In either case, the effects are either upon individuals who have been exposed, i.e. *somatic effects* , or upon their offspring as a result of damage to gonads, i.e. *genetic*, or *hereditary effects*. The most important somatic effects are the induction of malignant diseases, particularly leukaemia and cancer of the bone, lung, thyroid and breast – all of which also occur naturally. (Leukaemia occurs 'naturally' at a rate of ~1 in 20 000 per year). These diseases have been studied experimentally with a variety of mammalian species, but for man the only reliable data are obviously those obtained on himself. There are, however, a number of difficulties in interpreting such few data as are available. It has been shown that it is essential for irradiated populations to be studied over a considerable period of time, amounting to several decades. This requires that throughout the study period there is a consistency in diagnosis and that all malignancies are recorded. The provision of data from a suitable control population is also vital and, for any comparison to be valid, both populations must be large enough for the rates of occurrence of particular malignancies to be sufficiently high for an excess over that expected in the control population to be statistically significant. One of the difficulties here is that changes have apparently occurred in the 'natural' incidence of cancers in different populations during the past few decades, some of which have possibly resulted from the presence of other carcinogens in th environment.

A number of epidemiological surveys have demonstrated irradiation effects but the data have been of very limited value because it has not been possible to estimate the absorbed doses received by the individuals. In fact, with the exception of experiences gained in the use of irradiation in medicine, the majority of data have been obtained from circumstances in which accurate estimates of absorbed dose were not available. With regard to the very important studies on the survivors of Hiroshima and Nagasaki, although much effort has been made in estimating absorbed doses by knowledge of the distances individuals were from the hypocentres of the explosions, comparisons have been complicated because of the different principal types of radiations in the two events. In Hiroshima some 23 to 35% of the estimated kinetic energy released in material (or *kerma*) was attributed to neutrons, whereas at Nagasaki this percentage was estimated to be less than 2%. It is therefore difficult to compare estimates of the induction of malignant diseases per unit of absorbed dose at the two sites because, as was noted in chapter 1, radiations of high LET are more harmful than those of low LET per unit of absorbed dose. In addition, the RBE may vary for different types of effect and may also vary with the dose levels at which any comparisons are made. In view of all these qualifications it may reasonably be asked if any useful data can be derived at all. The answer is that very useful data have been obtained, as we shall see, but it is necessary to point out that unlike data obtained under controlled laboratory conditions, which may be verified by repeated experimentation, these epidemiological data have not been so derived and, one sincerely hopes, the circumstances from which they were derived will never be repeated.

A survey of the survivors of Hiroshima and Nagaski was started in 1950, and information obtained on their whereabouts at the time of the explosion. From such data, including allowances for shielding by buildings, etc, estimates have been derived of the radiation field to which the survivors had been exposed. The quantity used to specify the radiation field is the tissue kerma in free air (K) which, like absorbed dose, is measured in grays (or rads). Of the data obtained in this way, of particular value are those relating to the incidence of leukaemia. It has been found that leukaemia is induced by radiation with a mean latency period to diagnosis of about 10 years. The numbers which have occurred at Hiroshima and Nagasaki from 1950 to 1972 are given

TABLE 3.2. *Excess mortality from leukaemia in Hiroshima and Nagasaki from 1950 to 1972. The values are compared with the <0·09 Gy (<9 rad) group.*

Location	Dose group kerma in Gy (rad)		Observed cases	Expected cases	Excess	Excess rate per unit kerma 10^{-4} Gy^{-1} (10^{-6} rad^{-1})	90% confidence limits
Hiroshima	0·1–0·49	(10–49)	17	9·2	7·8	37	5–80
	0·5–0·99	(50–99)	7	2·2	4·8	29	7–67
	1·0–1·99	(100–199)	12	1·5	10·5	51	26–88
	>2	(>200)	28	1·3	26·7	57	39–79
Nagasaki	0·1–0·49	(10–49)	2	3·5	–1·5	–22	neg.–47
	0·5–0·99	(50–99)	0	1·2	–1·2	–15	neg.–24
	1·0–1·99	(100–199)	3	1·2	1·8	11	neg.–41
	>2	(>200)	15	1·2	13·8	35	20–52

Data from the Report to the United Nations Scientific Committee on the Effects of Atomic Radiation, 1977, General Assembly document 32 Session, Supplement No. 40 (A/32/40) (New York: United Nations).

in table 3.2. At first glance they may not seem to be all that useful. The mortality rates at Hiroshima in the range of 0·1 to 0·5 Gy (10 to 50 rad), and higher, do allow rates of leukaemia induction to be estimated; but at Nagasaki the lower dose groups are obviously too imprecise to be of value. If, however, estimates are based on the combined higher dose groups, the data are significant. For example, at Nagasaki the combined excess deaths at dose levels above 0·1 Gy (10 rad), over those expected by comparison with survivors who received less than this dose, was 13 (with 90% confidence limits of 6 and 22). The corresponding number of excess deaths at Hiroshima is much higher, at 50 (with 90% confidence limits of 37 and 65); this is because of the much higher neutron component at Hiroshima. For leukaemia induction such differences can be allowed for, however, with some degree of refinement, and if the data are recalculated in such a way as to allow for the different components of the radiation in the two cities, one can derive excess rates per 10^{-4} Gy^{-1} $(10^{-6}$ $rad^{-1})$ of 31 (with 90% confidence limits of 22 and 39) for Hiroshima, and 30 (with 90% confidence limits of 10 and 55) for Nagasaki. One may well ask, of course, if the number of cases observed in survivors who received doses of less than 0·1 Gy (10 rad) is different from what one might have expected. A comparison with people who were not in either city at the time has shown that the numbers of deaths from leukaemia in this group were higher ($P = 0.065$), and that they were also higher than expected on the basis of Japanese National Statistics ($P = 0.004$). Unfortunately neither of these comparisons really allows estimates to be made of the frequency with which leukaemia is induced at low absorbed doses.

There are other sources of data on radiation induced leukaemia, notably those resulting from pelvic X-irradiation and from radiotherapeutic treatment for ankylosing spondylitis. The data agree remarkably well and allow an overall estimate of the rate of leukaemia induction per unit of absorbed dose to be made. This estimate, for low LET radiation on a population of all ages, is currently 15 to 25×10^{-4} Gy^{-1} (15 to 25×10^{-6} rad^{-1}). Such a calculation is obviously very valuable but it is still important to note that it has largely been derived from absorbed radiation doses of over 1 Gy (100 rad).

Although induction of leukaemia in the Japanese A-bomb survivors has been the most single dramatic late irradiation effect, a number of other cancers have also been induced, the majority with longer latent periods. These data have been summarized in table 3.3. As for leukaemia, data are also available from a variety of sources. Radiation-induced cancer of the thyroid has resulted from patients

TABLE 3.3. *Excess mortality, compared with <0.09 Gy (<9 rad) group, from all causes in Hiroshima and Nagasaki from 1950 to 1972.*

Cause of death	Observed cases	Expected cases	Excess	Excess rate per unit kerma 10^{-4} Gy^{-1} (10^{-6} rad^{-1})	90% confidence limits
Leukaemia	84	21·0	63	36	27–46
Cancers of:					
stomach	417	411·8	5	3	–20–28
other digestive organs	261	255·3	6	3	–13–20
trachea, bronchus, lung	100	76·2	24	13	3–24
other respiratory organs	29	26·0	3	2	– 5–10
breast	37	24·0	13	7	1–14
cervix and uterus	86	77·5	8	4	– 4–14
other malignant neoplasms	145	117·0	28	16	3–30
unspecified neoplasms	53	52·6	0·4	0·2	
All diseases except neoplasms	3970	3947·7	22	13	–60–88

Data from the Report to the United Nations Scientific Committee on the Effects of Atomic Radiation, 1977, General Assembly document 32 Session, Supplement No. 40 (A/32/40) (New York: United Nations).

therapeutically irradiated from external sources, and from the internal administration of ^{131}I, to treat thymic enlargement. Cancers of the breast have, in the past, been induced by multiple fluoroscopic examinations during the treatment of pulmonary tuberculosis and from radiotherapy of the breast area for the treatment of acute mastitis. Other cancers, such as that of the lung in uranium miners exposed to high radon levels, and cancers resulting from a variety of early therapeutic treatments using ionizing radiations, have also been studied in detail, allowing some estimate to be made of their rates of induction per unit of absorbed dose. Some of these estimates, derived from the A-bomb survivors and all other sources, are given in table 3.4. The latent period of induction varies not only from cancer to cancer but frequently depends upon the age of the person at the time of irradiation.

Unfortunately, not all of the risk of induction estimates are derived from low-LET radiation data, i.e. from those which would be most applicable to environmental exposure. The human epidemiological estimates derived from whole-body neutron exposure at Hiroshima, from uranium miners exposed to radon daughters, from persons who have developed cancer of the liver resulting from injections of thorotrast (an X-ray contrast medium containing thorium) and data

TABLE 3.4.   *Estimates of the induction of some malignant diseases in relation to absorbed dose; not all would be fatal, depending upon treatment.*

Disease	Induction risk 10^{-4} Gy^{-1} (10^{-6} rad^{-1})
Leukaemia	15– 25
Thyroid cancer	50–150[†]
Breast cancer	
A-bomb data	30
fluoroscopic examination data	50–190
radiotherapy (bilateral) data	210[‡]
Lung cancer	25– 50
Bone cancer	2– ·5
Cancer of digestive organs	5– 10
Liver cancer	10– 20

[†] Within 25 years, equal numbers to be detected subsequently. About 1 in 10 would be fatal.
[‡] (R^{-1}).

Data from the Report to the United Nations Scientific Committee on the Effects of Atomic Radiation, 1977, General Assembly document 32 Session, Supplement No. 40 (A/32/40) (New York: United Nations).

derived from the irradiation of bone cells with ^{224}Ra and ^{226}Ra – all of these are largely derived for high-LET radiation. There is a need, therefore, for establishing a basis of comparison between high and low-LET radiation risk estimates. Bearing such difficulties in mind it is, nevertheless, possible to estimate the overall number of induced malignancies which would prove fatal from exposure to low-LET radiation in excess of 1 Gy (100 rad). Taking the leukaemia estimate as being 20×10^{-4} Gy^{-1} (20×10^{-6} rad^{-1}), the total risk, including leukaemia, is about five times higher, i.e. 100×10^{-4} Gy^{-1} (100×10^{-6} rad^{-1}).

All of the above estimates are for adults. Estimates for children would be higher, and estimates for the foetus exposed *in utero* considerably more so. Data on the latter have been obtained principally from mothers X-rayed while pregnant and therefore the estimates for induction are derived from much lower absorbed doses, in the range of 2 to 200 mGy (0·2 to 20 rad). The risk per unit absorbed dose for all fatal induced malignancies during the first 10 years of life, resulting from foetal irradiation, is estimated as being in the range of 200 to 250 $\times 10^{-4}$ Gy^{-1} (200 to 250 $\times 10^{-6}$ rad^{-1}), of which half would be due to leukaemia and one quarter to tumours of the nervous system.

One interesting and important feature which has emerged from the human epidemiological studies is that no evidence has been obtained to suggest that the induction rates of cancer, *per unit of absorbed dose*, to different tissues resulting from radionuclides incorporated into the body, differ from those resulting from external irradiation. In other respects, however, the human data have a number of obvious shortcomings and a very considerable effort has been expended on studies with other mammals. Such experimental studies have been of very great value in providing information on the relative carcinogenic effects of different types of radiation – notably the much greater effectiveness of alpha particles – and the many parameters which can alter such effects, such as the rates at which the doses are given, the fractionation of the dose, the age and sex of the animal, plus any other additive environmental agencies. Animal studies have also proved to be of great value in attempting to derive cancer risk estimates at different absorbed doses of radiation. It has been found that in some species, and for some cancers, the rates of induction are fairly proportional to the magnitude of the absorbed dose although, in a number of cases, the induction rate is relatively greater at larger doses.

Cancer is clearly the most important delayed cause of death, but there are several other important delayed effects of radiation. One somatic effect of particular relevance to the assessment of acceptable levels of exposure at work is the production of cataracts of the lens of the eye. The absorbed doses known to produce these vary from about 0·3 Gy (30 rad) for neutrons to about 5 Gy (500 rad) for beta or gamma radiation. A rather intriguing somatic effect of radiation in animals is an acceleration of the ageing process; that is, a loss of vitality coupled with both physical and mental deterioration. A lowering of resistance to infection has also been noted, and it is possible that such shortening of the life-span is, in fact, associated with effects on the immune · system. It appears that the incidence of auto-antibody formation naturally increases with age, and the insult of radiation may increase this rate of formation.

Apart from the induction of cancer, however, the effects of radiation which are of prime concern are those relating to genetics. There are, of course, a number of possible deleterious effects on the gonads which are not genetic. Human gonads appear to have a relatively low sensitivity to the induction of cancer but a decrease in the number of gametes of both male and female can result at relatively low absorbed doses. For example 0·25 Gy (25 rad) delivered at a high dose rate can depress, temporarily, the sperm count. It is the effects on the nuclear contents of the gametes themselves, particularly of the spermatogonia and oocytes, which are of greatest concern because any induced changes may be transmitted to the descendants of the persons irradiated. The nuclei of cells contain a quantity of DNA (deoxyribonucleic acid), a type of molecule which has the unique property of being able to make an exact copy of itself. Before a cell divides, its DNA becomes organized into paired structures, the chromosomes. In man there are 23 such pairs, i.e. 46 chromosomes altogether, each pair being very similar, with the important exception of the sex chromosomes. Before a cell divides, each chromosome divides, as the DNA duplicates itself, to form two chromatids. The two resultant cells each receive a chromatid from each chromosome, and thus both daughter cells resemble the original cell because they each have copies of the original 23 pairs of chromosomes. An exception occurs in the formation of the gametes.

In this case there is first of all a cell division in which two daughter cells are formed, each of which has only half the original number of chromosomes – although each chromosome consists of a rearrangement of two chromatids. The two daughter cells immediately divide again, normally, thus ultimately producing four gamete cells each of which has, in the case of man, 23 *single* chromosomes. Fertilization of one ovum by one sperm produces a cell, the zygote, which therefore has a full complement of 46 chromosomes.

Each chromosome is, in fact, an assemblage of discrete units of DNA called *genes*, the units being arranged in bead-like manner along the chromosome. In man the total number of different genes is estimated to be about 30 000. Radiations can cause structural damage to whole chromosomes resulting in their breakage and possible subsequent rejoining. This may result in changes to the sequence of the genes, effects known as *inversions* or *translocations*, or may even result in the absence or doubling of certain genes in that chromosome, effects referred to as *deletions* and *duplications*, respectively. Indeed the damage may be such as to result in a change of the number of chromosomes observed.

Alterations (mutations) to the genes on a chromosome are classified, for convenience, as being either *dominant* or *recessive*, depending upon the extent to which the effect is expressed in an offspring which inherits the mutated gene from one parent only. Thus a mutation which is fully dominant will have an effect if inherited from one parent only, whereas a fully recessive mutation has no effect unless the same mutated gene is received from both parents. The only exception is when one mutated gene is associated with the female chromosome. It was stated above that the chromosomes are paired with the exception of the sex chromosomes. Women have 23 matched pairs of chromosomes but in men the 23rd pair do not match. Instead of two large X chromosomes (XX) there is a large X and a small Y chromosome (XY). The Y chromosome carries the gene for maleness and is dominant over the X chromosome. Sperm may have either the X or the Y chromosome, whereas all ova have the X chromosome, and thus all resulting embryos receive at least one X chromosome. The relevance of this is that the X chromosome carries a number of genes which have nothing to do with characteristics of sex, whereas the Y chromosome does not.

Gene mutations and chromosome aberrations occur spontaneously in man; they are a source of considerable hardship. Chromosomal anomalies are frequent in spontaneously aborted zygotes – much higher than in live-born children – and their frequency of occurrence is dependent upon gestational age. Thus in one survey it has been shown that chromosomal anomalies of one kind or another were recorded in 66% of human zygotes which did not develop beyond 8 weeks, whereas this percentage fell to 23% for those aborted after 8 to 12 weeks development. The majority of these anomalies are in errors of chromosome number, for example having one extra chromosome homologous with one of the existing pairs (trisomy) or having three times the number of chromosomes in the unfertilized gamete (triploidy). Developmental arrests due to chromosomal anomalies in the first few weeks of pregnancy may be even greater, but such spontaneous abortions are frequently not recognized.

The number of children born with hereditary, or partially hereditary, defects and diseases, is surprisingly large. For every 1000 live-born children about 10 will be affected by a dominant genetic disease, 1 by a recessive genetic disease, 4 by a chromosomal disease and 90 by a variety of congenital malformations, multifactorial and irregularly inherited conditions – although not all of these 90 contain a mutational component. Almost all of the chromosomal disease is accounted for by Down's syndrome (mongolism). Many of the genetic diseases are part of the permanent genetic load carried by man. For example, in Britain the commonest disease due to a single pair of recessive genes is cystic fibrosis – a serious danger to life. As it is recessive, both parents must transmit the gene to the offspring for the disease to appear; in the population at large, about one person in twenty carries it. If two such persons marry, there is a one in four chance of their children developing the disease. Other human genetic disorders, such as achondroplasia (a form of dwarfism), are dominant and should therefore only be transmitted by an achondroplastic person – because posssssion of a dominant gene must automatically show its effect. But it is possible for normal people to have an achondroplastic child, as a result of a spontaneous mutation in either the ovum or the sperm of one of the parents. The mutant gene is now part of the child's genetic make-up, and his or her children will have a 50/50 chance of inheriting it. Data are continually being evaluated to assess the natural mutation rate of human genes, of which table 3.5 contains a selection. They have been divided into two groups; those which are sex-linked recessive mutations, i.e. carried by sex chromosomes, and those which are carried by the non-sex, or autosomal, chromosomes.

Very little quantitative information on the hereditary genetic effects of radiation in man is available. Fortunately children born to the survivors of the bombing in Hiroshima and Nagasaki have not had structural or numerical anomalies of the chromosomes significantly different from control groups. The mortality rate of children born to survivors has also failed to show, as yet, any significant effects of parental exposure. Recourse has therefore had to be made to animal studies, particularly on the mouse. An enormous effort has been expended on such studies, and will continue to be made. Two principal methods have been used to determine the effects. One is the '*direct*' method, by which the risks are expressed in terms of the expected frequencies of various types of genetic damage induced per unit of absorbed dose. The second method derives a *'doubling dose'*, in which estimates are made of the absorbed radiation doses which are required to double the occurrence of different types of genetic abnormalities which occur naturally. The expected effect of a given absorbed dose is then estimated on a proportional basis from the known natural frequencies of the abnormality in man and from the value assumed for the *'doubling dose'*. Using the 'direct' method, the total rate of induction of recessive mutations, per gamete, per unit of absorbed dose, is about 60×10^{-4} Gy^{-1} (60×10^{-6} rad^{-1}). The overall estimate of the induction of dominant effects is about 20×10^{-4} Gy^{-1} (20×10^{-6} rad^{-1}). Thus the genetic risk in the first generation following irradiation of

TABLE 3.5.   *Some estimated rates of mutation of human genes.*

Disease	Population studied	Number of mutants per million gametes
(A) Autosomal mutations		
Achondroplasia[1]	Northern Ireland	13
Aniridia[2]	Michigan, USA	3
Dystrophia myotonica[3]	Switzerland	11
Retinoblastoma[4]	England	6– 7
Acrocephalosyndactyly[5]	England	3
Osteogenesis imperfecta[6]	Sweden	7– 13
Tuberous sclerosis[7]	China	6
Neurofribromatosis[8]	Moscow, USSR	
Polyposis intestini[9]	Michigan, USA	13
Polycystic disease of kidney[10]	Denmark	65–120
Diaphyseal aclasis[11]	Munster, Germany	6– 9
(B) Sex-linked recessive mutations		
Haemophilia[12]	Denmark	32
Haemophilia A	Finland	32
Haemophilia B	Finland	2
Duchenne-type muscular dystrophy[13]	Wisconsin, USA	92

Data from the Report to the United Nations Scientific Committee on the Effects of Atomic Radiation, 1977, General Assembly document 32 Session, Supplement No. 40 (A/32/40) (New York: United Nations).

[1]A form of dwarfism. [2]Absence or defect of iris. [3]A degenerative disease. [4]Cancer of the retina. [5]Malformed head plus adhesion between fingers and/or toes. [6]Condition resulting in abnormal fragility of bones. [7]Tumours on certain areas of the brain. [8]Tumours arising from the fibrous tissues of nerves. [9]Formation of polyps on the mucous membranes of the intestine. [10]Formation of cysts in the kidney. [11]Superfluous bone formation. [12]Excessive haemorrhaging from trivial injuries. [13]A particular variety of a wasting disease of the muscle.

parents with 10 mGy (1 rad) is likely to be about 20 seriously affected
cases per million of live born; a further 2 to 10 cases per million would
be expected as a result of structural chromosome aberrations resulting
from the same absorbed dose. The 'doubling dose' method has shown
that the absorbed dose required to double the natural frequency of a
number of different forms of genetic abnormality is about 1 Gy (100
rad) for X-rays, beta and gamma radiation delivered at a low dose
rate. The increase of genetically determined diseases is therefore
unlikely to increase by more than 1% over the naturally occurring
rates per 10 mGy (1 rad) of absorbed dose. The breakdown of these
diseases into classes is given in table 3.6. Thus, in comparison to the
'direct' method, the 'doubling dose' method estimates that in the first
generation of children of a population exposed to 10 mGy (1 rad) at

TABLE 3.6.   *Estimated effect of 10 mGy (1 rad) of low-dose, low dose-rate,
low-LET irradiation, per human generation of one million live-born
individuals assuming a doubling-dose of 1 Gy (100 rad).*

Disease class	Natural incidence	Effect of 10 mGy (1 rad) per generation	
		First generation	Equilibrium
Autosomal dominant and X-linked diseases	10000	20	100
Recessive diseases	1100	Slight	Slow increase
Chromosomal diseases	4000	38	40
Other anomalies	90000	5	45
Total as a % of current incidence		0·06	0·18

Data from the Report to the United Nations Scientific Committee on the Effects of
Atomic Radiation, 1977, General Assembly document 32 Session, Supplement No. 40
(A/32/40) (New York: United Nations).

low dose rate there would be 63 genetic diseases, per million live born,
induced by irradiation. The total genetic damage expressed over all
generations is estimated to be ~185 per million per 10 mGy (1 rad), and
the same value applies for each generation after prolonged continuous
exposure.

In view of these calculations it might well be asked: what of the
populations living in areas with a higher than average natural
background level of irradiation? Such populations have, in fact, been
studied but so far have yielded inconclusive results. In the Brazilian
area of Espirito Santos (chapter 2) for example, a survey published in

1978 failed to detect any significant increase in sex ratio at birth, the occurrence of congenital anomalies, the numbers of pregnancy terminations, still-births, live-births, post-infant mortality in children, and fecundity and fertility of adults of populations receiving absorbed dose levels an order of magnitude greater than those of nearby 'control' populations. The study sample included more than 8000 couples. In contrast, a study of populations in the coastal area of Kerala, in India, has shown the frequency of Down's syndrome to be significantly higher, particularly for the maternal age group of 30 to 39 years, than that of control populations elsewhere in India. The frequency of Down's syndrome does vary very markedly from one population to another throughout the world, however, and these overall variations do not appear to be related to absorbed dose levels of irradiation. Clearly such epidemiological studies will always be difficult to interpret because even non-significant results, such as those of the Brazilian study, cannot be considered as not showing deleterious effects of low-level irradiation; they may simply imply that other causes of mortality, and genetic defects in live-born children, are sufficient to make negligible any effect of such irradiation. It is also important to remember that the types of genetic damage scored in the majority of studies – on either humans or other animals – do not usually include minor deleterious effects which may go unnoticed. There is no call for complacency, therefore, and further studies are continuing to be made. A recent report lists no less than 21 discrete areas of research in which further effort could provide useful data to increase our precision in genetic risk assessment.

3.3. *Standards for radiation protection*

From all of the evidence available to date it cannot be shown that there exists an absolutely safe dose of ionizing radiation for man; neither can it be shown that all radiation is harmful. Indeed, as we have already seen, there is no zero level of radiation. The greatest difficulty in devising standards for the protection of man – and other components of the environment – from excess ionizing radiations is the gap between the absorbed dose levels at which most effects are observed, that is above 1 Gy (100 rad), and the natural radiation background which gives a whole-body absorbed dose about four orders of magnitude lower. In addition, the majority of effects which have been studied in man derive from acute exposures, whereas the data which are most required are those relating to the potential effects resulting from chronic exposure to low levels of radiation over an

extended period of time. How should one bridge the gap? For the
dose/effect relationship, the easiest way out of the dilemma is to use
the data obtained with high dose levels, where some degree of dose/
effect relationship is observed, and then to extrapolate in a linear fashion
down to relatively normal background levels, but this is not necessarily
correct. A more typical biological response to stress is that which
follows a sigmoid curve in which, at low dose rates, the response to
stress incorporates some degree of tolerance via adaptation and repair
mechanisms. Both linear and sigmoid forms of response may have
threshold values, as illustrated in figure 3.1. It will be evident that at
low dose rates the linear form, without threshold, is the most conservative; it is this form which has, until recently, been used to set all
radiological standards for man.

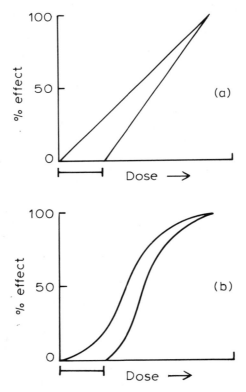

Figure 3.1.   Dose-response curves: (*a*) Linear and (*b*) sigmoid. Either form
may, or may not, have a threshold dose, below which there is no response
(⊢——⊣).

The basic responsibility for the provision of guidance, at an international level, on matters of radiological protection is that of the International Commission on Radiological Protection (ICRP) which has its origins in the setting up, in 1928, of the International X-ray and Radium Protection Commission. The present title was adopted in 1950. The Commission consists of members drawn from a number of countries. There are four committees, all of which routinely publish reports containing advice and recommendations to deal with the basic principles of radiation protection. The reports are by no means tablets of stone; as new data arise with regard to radiation effects such data are duly considered and, where appropriate, recommendations are revised accordingly. The amount of scientific literature addressed to the biological effects of radiation is increasing all the time and the important function of reviewing this literature has been undertaken, since 1955, by the United Nations Scientific Committee on the Effects of Atomic Radiation (UNSCEAR) whose most recent report (1977) has been the source of much of the data discussed in this chapter and in chapter 2. Another specialized agency of the United Nations, the International Atomic Energy Agency (IAEA) has, since 1956, been concerned with the practical applications of the ICRP recommendations. It is important to note that the ICRP data are recommendations only, and that their interpretation at a national level differs in some respects from one country to another.

It has always been assumed that exposure to ionizing radiations *over and above natural background levels* involves some degree of risk, and thus the question is raised as to what dose of absorbed radiation is acceptable to the individual, and to the public in general, in view of the benefits derived from such activities. Indeed the ICRP states that "no practice shall be adopted unless its introduction produces a positive net benefit". We will return to this interesting question in chapter 5, but for the present let us examine those doses which are currently recommended by the ICRP as maxima for 'occupational exposure', i.e. for adults exposed to controllable sources of radiation in the course of their work. These individuals are aware of the radiation hazards associated with their work and their working environment is registered, controlled, monitored and generally set out such that the recommended dose limitations are met. Personal dosimeters are also worn in order to provide a retrospective record of absorbed dose. Individuals are also subject to initial and regularly repeated medical examination.

In radiological protection use is made of the term *dose equivalent* (H), rather than absorbed dose (D). As discussed in chapter 1, the dose equivalent attempts to relate radiation exposure to biological effect on man by making use of the effective quality factors given in table 1.2 plus any other modifying factors considered to be influential – such as absorbed dose rate and fractionation. The dose equivalent is expressed in Sv (or rem).

A fundamental recommendation has been, until recently, that for occupational exposure an upper limit be set such that an individual should not accumulate a life-time whole-body dose greater than 50 $(N - 18)$ mSv $(5(N - 18)$ rem), where N is the individual's age in years. Accumulated exposures were assessed in time units of years and ICRP No. 9 (1966) further suggested that within this time period the absorbed dose should not exceed 30 mSv (3 rem) for any consecutive 13 week period, or 13 mSv (1·3 rem) for women of reproductive capacity. The *maximum permissible annual doses* to different parts of the body, from combined external and internal sources, also differed. Hands, forearms, feet and ankles were limited to 0·75 Sv (75 rem) per year; skin, bone and thyroid to 0·3 Sv (30 rem) per year; gonads and red bone-marrow to 50 mSv (5 rem) per year; and other single organs were limited to 0·15 Sv (15 rem) per year. All of these dose limitations were derived from linear, non-threshold, assumptions and have formed the basis for radiation standards all over the world.

More recently, ICRP No. 26 (1977) has revised these earlier recommendations. The new recommendations are notable for recognizing that for many of the possible effects resulting from radiation exposure one can only relate absorbed dose to the *probability* of the effect occurring. Two types of effect are therefore recognized: *stochastic* effects for which, like the hereditary effects and the induction of cancer, no threshold dose is assumed, but the probability of occurrence increases with dose; and *non-stochastic* effects for which, like lens cataract, the severity of the effect varies directly with the dose and thus allows the calculation of a 'threshold' value. Other non-stochastic effects are non-malignant damage to the skin, cell depletion in the bone-marrow causing haematological deficiencies and gonadal cell damage leading to impairment of fertility. The stochastic approach, therefore, recognizes the existence of a sigmoidal relationship which may be described mathematically by

$$E = aD + bD^2 \qquad (3.1)$$

where E denotes effect, D the absorbed dose and a and b are constants. The quadratic term bD^2 predominates at the high absorbed doses – above 1 Gy (100 rad) – and at high dose rates, of the order of 1 Gy min^{-1} (100 rad min^{-1}). At lower absorbed doses, and dose rates, the linear term aD predominates. The Commission justifies this change in approach by pointing out that the use of linear extrapolations from high doses, although sufficient for the assessment of upper limits of risk in comparison with an expected benefit, may not necessarily serve at lower doses. In fact the more cautious the linearity assumption, the more important it becomes to recognize that it may lead to an overestimate of the radiation risks; this, in turn, could therefore result in the choice of an alternative practice which could be more hazardous than that involving exposure to radiation.

The Commission also now recognizes that, as we have seen, the induction of malignant diseases is the predominant stochastic effect – a significant departure from earlier views. The risk factors for radiation protection purposes now recommended are summarized in table 3.7.

TABLE 3.7. *Risk factors (mortality) in relation to dose equivalent, compiled from ICRP No. 26 (1977).*

Effect	Risk factor, Sv^{-1} (rem^{-1})	
Leukaemia	2×10^{-3}	(2×10^{-5})
Bone cancer	5×10^{-4}	(5×10^{-6})
Lung cancer	2×10^{-3}	(2×10^{-5})
Thyroid cancer	5×10^{-4}	(5×10^{-6})
Breast cancer	$2 \cdot 5 \times 10^{-3}$	$(2 \cdot 5 \times 10^{-5})$
Other cancers	5×10^{-3}	(5×10^{-5})
Total stochastic risk from uniform whole-body irradiation	10^{-2}	$(\quad 10^{-4})$
Hereditary health risk in first two generations following irradiation of either parent	4×10^{-3}	(4×10^{-5})

Data derived from Annals of the ICRP, 1977, Publication No. 26, Recommendations of the International Commission on Radiological Protection, (Oxford: Pergamon Press); courtesy of the publishers.

The annual dose limit is still 50 mSv (5 rem), a limit which is expected to prevent the occurrence of non-stochastic effects, for which 0·5 Sv (50 rem) would suffice – with the exception of the lens of the eye, which would require a limit of 0·3 Sv (30 rem) – and also to limit the frequency of stochastic effects to an appropriately low level. But the quarterly limit, and the age-related formula given above, are no longer recommended. Limits for local tissue irradiation are set by 'weighting factors' (W_T) relative to the whole body. These are given in table 3.8. It is assumed that the safety of the worker exposed to radiation should be equally maintained whatever the manner of irradiation, that is whether the whole body is irradiated uniformly or particular organs are irradiated selectively. The weighting factors in table 3.8, for which the whole body is taken as 1·0, reflect the 'harm' attributable to irradiation of each, and on the assumption that interaction between exposure of one organ and the induction of fatal malignant, or severe genetic damage, in another will be minimal at low doses.

TABLE 3.8.   *Weighting factors for individuals, from ICRP No.26 (1977). The sum of the weighted dose equivalents must not exceed 50 mSv (5 rem).*

Tissues	Weighting factor (W_T)
Gonads	0·25
Breast	0·15
Red bone-marrow	0·12
Lung	0·12
Thyroid	0·03
Bone surfaces	0·03
Remainder	0·30
	$\Sigma\ 1\cdot00$

From *Annals of the ICRP*, 1977, Publication No. 26, Recommendations of the International Commission on Radiological Protection, (Oxford: Pergamon Press); courtesy of the publishers.

Limits for individual organs are no longer specifically recommended but implied limits can be derived by dividing the annual limit for the whole body by the weighting factor. For example, for the lung the implied limit would be 50 mSv (5 rem) divided by 0·12, giving 0·417 Sv (42 rem). To avoid the risk of non-stochastic effects the 0·5 Sv (50 rem) limit, with the exception of the lens as mentioned above, provides an additional overriding limitation – as for thyroid and bone surfaces.

Women of reproductive capacity who are occupationally exposed to radiation are unlikely to receive more than 5 mSv (0·5 rem) to the embryo during the first two months of a pregnancy which might remain unrecognized. Once the pregnancy has been diagnosed, however, a pregnant woman should continue to work *only* under conditions in which it is most unlikely that her annual exposure would exceed 30% of the usual dose equivalent limits.

Any regulations designed for occupational use must allow for exceptional circumstances. A planned special exposure – but not for women of reproductive capacity – is considered acceptable providing that it does not exceed twice the annual limit in any single event and, in a lifetime, five times this limit.

It is important to note that these recommended limits for occupational workers are not designed to eliminate any 'harm' due to irradiation. Drawing comparisons between harm in different industries is clearly very difficult, but taking death as an index the Commission (ICRP No. 27, 1977) estimates that *continuous* annual exposure of *every* worker at the limit of 50 mSv (5 rem) per year would result in a fatality rate equal to 340×10^{-6} y^{-1}. In fact the average annual exposure to occupational workers has commonly been found to be 6 mSv (0·6 rem). The calculated effect at this level of radiation, in occupations involving radiation exposure, is a total fatality rate equal to 45×10^{-6} y^{-1}, of which a quarter would be attributable to accidental causes not associated with radiation, half would be attributable to fatal malignancies, and the remainder would be attributable to genetic effects – making the assumption that half of the workforce consisted of women and that pregnancies occurred at normal rates in the working population. For comparison, the accidental death rate in the USA for 1972 was 96×10^{-6} y^{-1} in the manufacturing industries, 657×10^{-6} y^{-1} in agriculture, 710×10^{-6} y^{-1} in the construction industries, and 1000×10^{-6} y^{-1} in mining.

The ICRP has also now recommended a method by which one can allow for the different risks of mortality associated with the irradiation of different organs, plus a proportion of the hereditary effects (those which would arise in the first two generations), by using a single value – the *effective dose equivalent* (H_E). As an index of total health detriment, therefore, even the effective dose equivalent is not complete because it does not include hereditary effects in all subsequent generations, and does not allow for non-fatal somatic effects such as the majority of cases of thyroid or skin cancer. Nevertheless, it is a very useful quantity. It is calculated by

$$\Sigma_T W_T H_T$$

where W_T is a weighting factor as given in table 3.8, H_T is the dose equivalent in tissue T, and the summation is made over the same period of time in all tissues.

To assess the H_E therefore, all that is required is to assess the dose equivalent in each tissue from all sources, multiply by the respective weighting factor, and sum the results. In principal it does not matter whether the dose equivalent results from internal or external irradiation. In practice, however, it will usually be more convenient to assess the contributions from external and internal radiation separately. It can be imagined that meeting the requirements for external exposure is a fairly straightforward process. But what of internal exposure? Radionuclides may be inhaled or ingested; they will become concentrated in different organs and remain there for different periods of time. How can one assess this rate of exposure, over and above natural background, and set limits for it? The practice has been based on the concept of a single critical organ. It is a practice which is being radically revised by the ICRP, although it has formed the basis for internal exposure assessment throughout the world.

It has long been known that different elements are differently metabolized and differently distributed within the body. Individual atoms of an element are also not usually permanently assimilated into the body but are exchangeable, to a varying degree, throughout life. For example, a widely distributed element is sodium, which is present in body fluids, cells and in bone. The sodium of the body fluids and cells of soft tissues is available for rapid exchange with sodium absorbed daily from the gut. Approximately 40% of the total body sodium occurs in bone, but nearly half of this is not permanently incorporated, so that virtually three-quarters of the sodium body-burden is exchangeable. Both chlorine and potassium are also widely distributed elements which are fairly readily exchangeable. There are also elements which concentrate in some organs to a far greater extent than in others. Iodine, for example, although occurring in all tissues of the body at concentrations of <1 μg g^{-1} wet accumulates in the thyroid gland to concentrations of about 600 μg g^{-1} wet. Thus the thyroid, an organ which constitutes only ~0.03% body weight, contains up to 12 mg iodine, and it is estimated that the total body burden is only ~13 mg. Other elements concentrate in a number of particular organs. Iron, for instance, is associated with erythropoietic tissues – those concerned with the production and storage of red blood cells. The total body-burden of iron is about 4 g, 60% of which is bound in haemoglobin of the red blood cells and their precursors; a further 10% is contained in the liver and spleen and the remainder distributed throughout the body. Its metabolism is therefore rather complex; so also is that of calcium, which occurs in exchangeable forms in soft tissues, and even in bone, for although deposited where the bone is growing it may be remobilized at other sites. The chemical form of the element when ingested may also be important. When hydrogen is assimilated as part of a molecule of water it is rapidly distributed throughout the body and subsequently excreted; when it is assimilated as part of an organic molecule, however, it may be retained for very much longer periods, dependent upon the metabolism of the molecule of which it forms a part. It must also be realized that the concentrations of elements in the whole body, and in various organs, may change during life. The best example of this is potassium, which shows a variation in concentration with both age and sex. The variation reflects the proportion of adipose tissue, which is relatively free of potassium, present in different individuals; there is generally relatively more in females, and the proportion increases, in both sexes, with age.

In order to calculate the impact, and thus recommend a permissible rate of intake, of a radionuclide, it is therefore necessary to compile a number of relevant data. With regard to the radionuclide, its type of emitted radiation (α, β, γ), the energy of the radiation, and its physical half-life and decay scheme are all relevant. Of the biological data required, it is usually necessary to determine that fraction of the radionuclide – inhaled or ingested – which is actually assimilated, the particular organs in which it may become concentrated, the ratio of any such concentration to that of the whole-body content, and the rate at which the radionuclide may be execreted, both from a particular organ and from the body as a whole. It has, in fact, as was mentioned above, been customary to recognize a particular part of the body which, for a given radionuclide, is of the greatest importance on account of the dose that it receives, its sensitivity to radiation, and the importance to health of any damage that results. Where definable, this organ (or organs) has been termed the *critical organ*.

In order to recommend maximum exposure limits it is obviously necessary, having taken account of the numerous variables, to generalize to a large extent, and in such a way as not to jeopardize those who differ markedly from the norm. The first question is obviously: what constitutes an average man or woman? The ICRP has gone to considerable lengths to compile and interpret an enormous amount of information on the constitution and variability of *Homo sapiens* in order to arrive at a standard man. Their most recent report on the subject, ICRP No. 23 (1975), summarizes available data to derive a *"Reference Man* for the Purposes of Radiation Protection". He is a most intersesting specimen and worth examining in some detail.

Reference Man comes in two sexes, the male having a total body weight of 70 kg wet and the female one of 58 kg wet; their heights are 170 and 160 cm, respectively. Reference Man is considered to be between 20 and 30 years of age; he is a Caucasian, lives in a climate with an average temperature of 10° to 20° C and is Western European or North American in habitat and custom. Part of the reason for Reference Man being so racially and culturally defined is because most of the data avaible relate to studies in these areas. Not that the data are intended, even here, to describe an 'average' individual of a specified population group; and they do not necessarily represent data which would be obtained by taking a random sample of the population. The important

TABLE 3.9. *Weights (wet) of principal organs and tissues of Reference Man (Total weight ♂ 70 kg: ♀ 58 kg).*

Tissue or organ	Weight (g)		% total body weight	
	♂	♀	♂	♀
Skeletal muscle†	28000	17000	40·0	29·3
Total skeleton†	10000	6800	14·3	11·7
cortical bone	4000		5·7	
trabecular bone	1000		1·4	
red marrow	1500		2·1	
yellow marrow	1500		2·1	
cartilage	1100		1·6	
Subcutaneous adipose tissue†	7500	13000	10·7	22·4
Other separable adipose tissue†	5000	4000	7·1	6·9
Skin†	2600	1790	3·7	3·1
Gastrointestinal tract (GI)†	1200	1100	1·7	1·9
oesophagus	40	30	0·06	0·05
stomach	150	140	0·2	0·2
intestine	1000	950	1·4	1·6
Contents of GI tract†	1005		1·4	
Liver†	1800	1400	2·6	2·4
Spleen†	180	150	0·3	0·3
Pancreas†	100	85	0·1	0·1
Lungs†	1000	800	1·4	1·4
Kidneys†	310	275	0·4	0·5
Testes†	35		0·05	
Ovaries†		11		0·02
Uterus†		80		0·14
Brain†	1400	1200	2·0	2·1
Spinal cord†	30	28	0·04	0·05
Heart†	330	240	0·5	0·4
Total blood‡	5500	4100	7·9	7·1
plasma	3100	2600	4·4	4·5
erythrocytes	2400	1500	3·4	2·6

† Used in total body calculations.
‡The weights of all organs except the heart include blood vessels etc., blood and other fluids. The heart is weight without blood in the chambers.

From ICRP Publication No. 23, 1975, Report of the Task Group on Reference Man (Oxford: Pergamon Press); courtesy of the publishers.

aspect of the Reference Man concept, for which all of the parameters and characteristics are so precisely defined, is that there is a standard, a basis, for the procedure of dose estimation. It remains for the various national or regional organizations which implement the control of radiation exposure locally to determine what modifications may be necessary for a particular population at risk.

The gross anatomical data relating to Reference Man are given in table 3.9. It is noteworthy that a much greater percentage of the body

TABLE 3.10. *Some physiological parameters of Reference Man.*

Dietary intake (g d⁻¹)	
as: protein	95
carbohydrates	390
fat	120
as: carbon	300
hydrogen	350
oxygen	2600
Faecal components (g d⁻¹)	
as: water	105
solids	30
ash	17
fats	5
as: carbon	7
hydrogen	13
oxygen	100
Water (fluid) intake (ml d⁻¹)	1950
Water intake in food (ml d⁻¹)	700
Water intake by oxidation of food (ml d⁻¹)	350
Water loss (ml d⁻¹) in urine	1400
in faeces	100
other	1500
Lung capacity (total)†	5·6 l
Functional residual capacity‡	2·2 l
Vital capacity§	4·3 l
Air breathed (l) 16 h awake (light work)	19200
8 h asleep	3600
Energy expenditure (kcal d⁻¹)	3000

† Volume of gas contained in the lungs after maximal inspiration.
‡ Volume of gas remaining in the lungs after normal expiration.
§ Volume of gas remaining in the lungs after maximal expiration.

From ICRP Publication No. 23, 1975, Report of the Task Group on Reference Man (Oxford: Pergamon Press); courtesy of the publishers.

weight of the male consists of skeletal muscle and bone, and that a much greater percentage of the female consists of subcutaneous fat. The female, however, has a slightly larger relative brain weight! A number of physiological parameters are summarized in table 3.10, for the male only. The potential variability here is considerable, even allowing for the limits of defining Reference Man as described above. Dietary intakes are clearly very variable, even with regard to the total daily food intake; and fluid intake will also differ markedly from one individual to another. As for the anatomical values, however, it must again be stressed that these are not considered to be 'average' values, but 'reference' values against which any other data – in so far as they affect the radiological consequences – can be compared. Also the amount of air breathed per day obviously depends on the effort exerted, and the 'fitness' of the individual. The values given in table 3.10 assume that the amount of air breathed during an 8-h working period ('light' work) is the same as that for an 8-h period of non-occupational activity.

TABLE 3.11. *Total body content of twenty major elements in Reference Man. (Total weight 70 kg adult male.)*

Element	Amount (g)	% of total body weight
Oxygen	43000	61
Carbon	16000	23
Hydrogen	7000	10
Nitrogen	1800	2·6
Calcium	1000	1·4
Phosphorus	780	1·1
Sulphur	140	0·2
Potassium	140	0·2
Sodium	100	0·14
Chlorine	95	0·12
Magnesium	19	0·027
Silicon	18	0·026
Iron	4.2	0.006
Fluorine	2·6	0·0037
Zinc	2·3	0·0033
Rubidium	0·32	0·00046
Strontium	0·32	0·00046
Bromine	0·20	0·00029
Lead	0·12	0·00017
Copper	0.072	0.00010

From ICRP Publication No. 23, 1975, Report of the Task Group on Reference Man, (Oxford: Pergamon Press); courtesy of the publishers.

Of special interest with regard to radiological protection is the chemical composition of Reference Man. Any general compilation of chemical data would immediately be beset by all of the problems of the accuracy, and intercomparability, of analyses made by different investigators. Even with the most modern equipment there are frequently considerable differences in the accuracy of chemical analyses because of contamination. It is fortunate, therefore, that the data used by ICRP for the most common elements are based on the analyses of 150 adults – victims of accidental death – which were made by the same laboratory. Thus they have a high degree of internal consistency and, in general, compare very well with values reported by other investigators. The twenty major elements are listed in table 3.11. This, of course, is merely a static picture; it demonstrates that over 96% of the human body is made up of oxygen, carbon, hydrogen and nitrogen. In contrast, table 3.12. attempts a more dynamic approach by estimating the values of daily intake and loss for a number of these elements. There is one fundamental observation to be made with regard to this table – apart from the fact that not all of the values neatly balance – and that is that the relative daily intake, and loss, of an element does not necessarily compare with its relative abundance in the human body (table 3.11). Thus potassium is absorbed and excreted daily in quantities much greater than calcium, even though the total body content of the latter is greater than that of potassium by almost an order of magnitude. This is because, while maintaining a total amount within the body, the rates at which many elements are continuously replaced differ from one element to another. Thus if one hundred 'labelled' atoms of an element are present within the body at time zero, these atoms will gradually be replaced and the number of 'labelled' atoms present continually reduced. This loss of the 'labelled' atoms can frequently be expressed by an exponential model in a manner analogous to that of the reduction in number of radioactive atoms present in a sample, as described in chapter 1. This permits the concept of a *biological half-time* for an element – analogous to the physical half-life of a radionuclide – such that a biological half-time of 20 days tells one that the number of 'labelled' atoms present at time zero has been reduced after 20 days, by 50%. If these 'labelled' atoms were radioactive, that is they were, for example, atoms of ^{131}I, after 20 days their number would not only be reduced by biological replacement, their number would also be reduced because of the radioactive decay of the atoms themselves. This combined effect is termed the *effective half-time*. The three half-times are related by

$$T_e = \frac{T_b \times T_p}{T_b + T_p} \tag{3.2}$$

TABLE 3.12. Model values for daily balance of some elements in Reference Man.

Element	Intake		Losses			Units
	Food and fluids	Airborne	Urine	Faecal	Other	
Oxygen ♂	2600	920	1300	100	2060	g
Oxygen ♀	1800	640	1100	90	1410	g
Carbon	300		5	7	288	g
Hydrogen	350		160	13	177	g
Nitrogen ♂	16		15	1·5	0·3	g
Nitrogen ♀	13		13	1·3	0·3	g
Calcium	1·1		0·18	0·74	0·03–0·15	g
Potassium	3·3		2·8	0·36	0·13	g
Iron ♂	16	0·03	0·25	15	0·51	mg
Iron ♀	12	0·03	0·20	11	0·51 + 0·6†	mg
Zinc	13	<0·1	0·5	11	0·8 + 1·0† ♀	mg
Rubidium	2·2		1·9	0·3	0·05	mg
Strontium	1·9		0·34	1·5	0·02	mg
Manganese	3·7	0·002	0·03	3·6	0·04	mg
Iodine	200	0·5–35	170	20	8·3	µg
Cobalt	300	< 0·1	200	90	6·4	µg
Caesium	10	0·025	9	< 1	trace	µg
Thorium	3		0·1	2·9		µg
Uranium	1·9		0·05–0·5	1·6	0·02	µg

† Menstrual loss.

where T_e is the effective half-time, T_b the biological half-time and T_p the physical half-life of the radionuclide.

In order to establish the amount of a radionuclide which can be safely assimilated by the body, additional physiological data need to be determined for each element: the fraction of the nuclide which reaches the blood, the fraction of the nuclide in the blood which reaches a particular organ, how this relates to the total amount in the body, and so on. All of these have been compiled by the ICRP and in one of their earliest reports (ICRP No.2, 1959), using an earlier version of the Reference Man discussed above, the Committee set out in detail recommended *maximum permissible body burdens* (*q*) for about 240 radionuclides, and recommended *maximum permissible concentrations* (MPC) of these radionuclides in water $(MPC)_w$ and in the air $(MPC)_a$ for either a 40-h working week for 50 weeks a year, or for continuous exposure, i.e. a 168-h week. The daily intake of water used in calculating the $(MPC)_w$ includes the water content of food. Where the nuclide is accumulated solely from contaminated food, it has usually been the practice to estimate a maximum oral intake, based on the MPC of that nuclide in water at a total daily water intake of 2·2 l, or 1·1 l in an 8-h working day. (The New Reference Man, ICRP No. 23, drinks slightly more!)

It was stated above that the recommendations are set to allow for combined external and internal exposure. The MPC values assume no external exposure, and thus where such exposure exists the MPC values must be reduced by the factor $(D - E/D)$, where D is the maximum dose permitted to an organ and E is the dose received from external irradiation. The MPC values were calculated on the basis that such a rate of intake would result in a dose to the critical organ equal to the maximum permissible dose rate after a period of 50 years, and would also not exceed the annual permissible dose to that organ – as stated in ICRP No.9 and discussed earlier. There are several nuclides which, because of their extremely long biological half-lives, are not at an equilibrium after 50 years: included in this list are ^{90}Sr, [226]Ra, [238]Pu, [239]Pu, [240]Pu and [241]Am. Thus, in theory, if *occupational* exposure continued beyond 50 years, the dose rate would continue to rise, and would exceed the maximum permitted. It was considered very unlikely, however, that such prolonged occupational exposure would occur.

The accumulation of a radionuclide into a critical organ, as opposed to the whole body, is described by

$$\frac{d\,(qf)}{dt} = P - \lambda\,(qf) \qquad (3.3)$$

TABLE 3.13. *Maximum permissible concentrations of some selected radionuclides in air and in water for occupational exposure* ($1\ \mu$Ci $= 3.7 \times 10^4$ Bq).

Radionuclide and type of decay		Critical organ	Maximum permissible concentrations (μCi cm^{-3})			
			For 40-h week		For 168-h week	
			$(MPC)_w$	$(MPC)_a$	$(MPC)_w$	$(MPC)_a$
^{3}H (β^-)	sol.	Body tissue	0·1	5×10^{-6}	0·03	2×10^{-6}
^{14}C (β^-)CO_2	sol.	Fat	0·02	4×10^{-6}	8×10^{-3}	10^{-6}
^{32}P (β^-)	sol.	Bone	5×10^{-4}	7×10^{-8}	2×10^{-4}	2×10^{-8}
	insol.	Lung		8×10^{-8}		3×10^{-8}
		GI(LLI)	7×10^{-4}	10^{-7}	2×10^{-4}	4×10^{-8}
^{35}S (β^-)	sol.	Testis	2×10^{-3}	3×10^{-7}	6×10^{-4}	9×10^{-8}
	insol.	Lung		3×10^{-7}		9×10^{-8}
		GI(LLI)	8×10^{-3}	10^{-6}	3×10^{-3}	5×10^{-7}
^{41}Ar (β^-, γ)	sol.	Total body		2×10^{-6}		4×10^{-7}
^{54}Mn (γ)	sol.	GI(LLI)	4×10^{-3}	8×10^{-7}	10^{-3}	3×10^{-7}
		Liver	0·01	4×10^{-7}	4×10^{-3}	3×10^{-7}
	insol.	Lung		4×10^{-8}		10^{-8}
		GI(LLI)	3×10^{-3}	6×10^{-7}	10^{-3}	2×10^{-7}
^{59}Fe (β^-, γ)	sol.	GI(LLI)	2×10^{-3}	4×10^{-7}	6×10^{-4}	10^{-7}
		Spleen	4×10^{-3}	10^{-7}	10^{-3}	5×10^{-8}
	insol.	Lung		5×10^{-8}		2×10^{-8}
		GI(LLI)	2×10^{-3}	3×10^{-7}	5×10^{-4}	9×10^{-8}
^{60}Co (β^-, γ)	sol.	GI(LLI)	10^{-3}	3×10^{-7}	5×10^{-4}	10^{-7}
	insol.	Lung		9×10^{-9}		3×10^{-9}
	insol.	GI(LLI)	10^{-3}	2×10^{-7}	3×10^{-4}	6×10^{-8}
^{65}Zn (β^+, γ)	sol.	Total body	3×10^{-3}	10^{-7}	10^{-3}	4×10^{-8}
		Prostate	4×10^{-3}	10^{-7}	10^{-3}	4×10^{-8}
		Liver	4×10^{-3}	10^{-7}	10^{-3}	5×10^{-8}
		Liver		6×10^{-8}		2×10^{-8}

Isotope		Organ				
^{85}Kr (β^-)	sol.	Total body		10^{-5}		3×10^{-6}
^{90}Sr (β^-)	insol.	Bone	10^{-5}	10^{-9}	4×10^{-6}	4×10^{-10}
		Lung	10^{-3}	5×10^{-9}	4×10^{-4}	2×10^{-9}
^{95}Zr (β^-, γ)	sol.	GI(LLI)	10^{-3}	2×10^{-7}	6×10^{-4}	6×10^{-8}
		GI(LLI)	2×10^{-3}	4×10^{-7}	1	10^{-7}
	insol.	Total body	3	10^{-7}		4×10^{-8}
^{95}Nb (β^-, γ)	sol.	Lung	2×10^{-3}	3×10^{-8}	6×10^{-4}	10^{-8}
		GI(LLI)	3×10^{-3}	3×10^{-7}	10^{-3}	10^{-7}
	insol.	GI(LLI)	10	6×10^{-7}	4	10^{-7}
		Total body	3×10^{-3}	5×10^{-7}		2×10^{-7}
^{99}Tc (β^-)	sol.	Lung	3×10^{-3}	5×10^{-7}		3×10^{-8}
		GI(LLI)	0.01	2×10^{-6}	10^{-3}	2×10^{-7}
	insol.	GI(LLI)	5×10^{-3}	6×10^{-8}	3×10^{-3}	7×10^{-7}
^{106}Ru (β^-, γ)	sol.	Lung	4×10^{-4}	8×10^{-7}	10^{-4}	2×10^{-8}
		GI(LLI)	3×10^{-4}	8×10^{-8}	2×10^{-3}	3×10^{-7}
	insol.	GI(LLI)		6×10^{-9}		3×10^{-8}
^{129}I (β^-, γ)	sol.	Lung	3×10^{-4}	6×10^{-8}	10^{-4}	2×10^{-9}
		Thyroid	10^{-5}	2×10^{-9}		2×10^{-8}
	insol.	Lung		7×10^{-8}		6×10^{-10}
^{131}I (β^-, γ)	sol.	GI(LLI)	6×10^{-3}	10^{-6}	2×10^{-3}	2×10^{-8}
		Thyroid	6×10^{-5}	9×10^{-9}	2×10^{-5}	4×10^{-7}
	insol.	Lung		3×10^{-7}		3×10^{-9}
^{133}Xe (γ)	sol.	GI(LLI)	2×10^{-3}	3×10^{-7}	6×10^{-4}	10^{-7}
		Total body		10^{-5}		10^{-7}
^{137}Cs (β^-, γ)	sol.	Total body	4×10^{-4}	6×10^{-8}	2×10^{-4}	3×10^{-6}
		Liver	5×10^{-4}	8×10^{-8}	2×10^{-4}	2×10^{-8}
		Spleen	6×10^{-4}	9×10^{-8}	2×10^{-4}	3×10^{-8}
		Muscle	7×10^{-4}	10^{-7}	2×10^{-4}	3×10^{-8}
		Lung		$\cdot 10^{-8}$		4×10^{-9}
	insol.	Gi(LLI)	10^{-3}	2×10^{-7}	4×10^{-4}	5×10^{-9}
						8×10^{-8}

TABLE 3.13. *Continued*

Radionuclide and type of decay		Critical organ	Maximum permissible concentrations (μCi cm^{-3})			
			For 40-h week		For 168-h week	
			$(MPC)_w$	$(MPC)_a$	$(MPC)_w$	$(MPC)_a$
^{144}Ce (β^-, γ)	sol.	GI(LLI)	3×10^{-4}	8×10^{-8}	10^{-4}	3×10^{-8}
		Bone	0.2	10^{-8}	0.08	3×10^{-9}
		Liver	0.3	10^{-8}	0.1	4×10^{-9}
	insol.	Lung		6×10^{-9}		2×10^{-9}
^{210}Po (α)	sol.	GI(LLI)	3×10^{-4}	6×10^{-8}	10^{-4}	2×10^{-8}
		Spleen	2×10^{-5}	5×10^{-10}	7×10^{-6}	2×10^{-10}
		Kidney	2×10^{-5}	5×10^{-10}	8×10^{-6}	2×10^{-10}
	insol.	Lung		2×10^{-10}		7×10^{-11}
^{235}U (α, β^-, γ)	sol.	GI(LLI)	8×10^{-4}	2×10^{-7}	3×10^{-4}	5×10^{-8}
		Kidney	10^{-4}	5×10^{-10}	4×10^{-5}	2×10^{-10}
		Bone	10^{-4}	6×10^{-10}	5×10^{-5}	2×10^{-10}
	insol.	Lung		10^{-10}		4×10^{-11}
^{238}U (α, γ)	sol.	GI(LLI)	8×10^{-4}	10^{-7}	3×10^{-4}	5×10^{-8}
		Kidney	2×10^{-5}	7×10^{-11}	6×10^{-6}	3×10^{-11}
	insol.	Lung		10^{-10}		5×10^{-11}
^{239}Pu (α, γ)	sol.	GI(LLI)	10^{-3}	2×10^{-7}	4×10^{-4}	6×10^{-8}
		Bone	10^{-4}	2×10^{-12}	5×10^{-5}	6×10^{-13}
	insol.	Lung		4×10^{-11}		10^{-11}
^{241}Am (α, γ)	sol.	GI(LLI)	8×10^{-4}	2×10^{-7}	3×10^{-4}	5×10^{-8}
		Kidney	10^{-4}	6×10^{-12}	4×10^{-5}	2×10^{-12}
		Bone	10^{-4}	6×10^{-12}	5×10^{-5}	2×10^{-12}
	insol.	Lung		10^{-10}		4×10^{-11}
		GI(LLI)	8×10^{-4}	10^{-7}	3×10^{-4}	5×10^{-8}

From ICRP Publication No. 2, 1959, Recommendations of the International Commission on Radiological Protection, Oxford: Pergamon where P is the rate of uptake of a radionuclide by the critical organ (Bq day^{-1} or pCi day^{-1}); qf is the burden of the radionuclide in the organ; λ is $0.693/T_e$ and t is the period of exposure, set at 50 years. It follows, therefore, that when $t = 0$, $qf = 0$ and thus

$$qf = \frac{P(1 - e^{-\lambda t})}{\lambda} \qquad (3.4)$$

There are, in fact, numerous basic equations used for the calculation of maximum permissible body burdens; when the radionuclide decays into a whole series of daughters the mathematics become rather complicated. Of course not all radionuclides are absorbed into the body. Many are ingested, passed along the gastrointestinal tract, and then excreted. A specific mathematical model has therefore been developed in order to calculate the maximum permissible absorbed dose to various portions of the gut, taking into account the physical aspects of the radionuclide's decay and the time spent in different portions of the gastrointestinal tract.

Not all radionuclide metabolism can be represented by an exponential assumption of their rates of loss. The notable exception is the rate at which a number of radionuclides are eliminated from bone, for which the data are better described by a power function model of the form

$$R_t = At^{-n} \qquad (3.5)$$

where R_t is the fraction of the deposited radionuclide remaining in an organ at time t, A is the fraction remaining after unit time and n is a constant. However, when maximum permissible concentrations are calculated using a power function model, because it implies that the *rate* of excretion is continually increasing, the drived values are greater than those calculated on the basis of an exponential model. The exponential model is preferred, therefore, not only for its convenience –because it can be readily related to the physical half-life of a radionuclide – but because of its conservatism. It should also be noted that a single exponential rate of loss is likely to be an extreme simplification of the actual processes involved.

There is a considerable amount of data relating to the deposition of radium in the human body, and the maximum permissible body burden of ^{226}Ra and its daughter products corresponds to an average dose rate to bone of 56 mSv week^{-1} (0.56 rem week^{-1}). The maximum permissible body-burdens of other bone-seeking radionuclides, such as those of strontium, actinium, thorium, protactinium, neptunium and plutonium, are therefore calculated from this basic value. All of these elements have biological half-times in bone of between 10^4 and 10^5 days and thus are, effectively, permanently incorporated and not excreted. A relative damage factor is derived which takes into consideration the fact that some radionuclides do greater damage to bone, per unit of absorbed dose, than others.

For airborne radioactivity, the lung is considered both as a route of entry into the human body and as a critical organ in itself. Gaseous radionuclides may be absorbed across the capillary walls of the blood vessels of the lung. Particulate radionuclides which are inhaled may also be absorbed if they are soluble, but 'insoluble' particulates can be deposited within the smaller branches of the respiratory tract and remain there for long periods of time. Eventually these particulates may be removed by exhalation, they may slowly dissolve, or they may be removed by phagocytosis. For soluble particles it is assumed that 25% are exhaled, 50% are deposited in the upper respiratory tract and are eventually swallowed, and 25% become dissolved and absorbed into the body fluids via the lung. For insoluble particles it is assumed that 25% are exhaled, 62·5% eventually swallowed, and that 12·5% are deposited deep within the respiratory tract, and remain there with a biological half-time of 120 days.

The deposition of alpha-emitting isotopes of one particular element, plutonium, in the lung has recently been the subject of some controversy; this has centred around the estimation of damage which it is assumed could result from such an alpha-emitting, particulate, radionuclide which is evenly deposited within the lung, relative to that which could result if the same quantity was not so distributed but localized into 'hot spots'. The fears which were initially raised do not appear to be substantiated, however, and although it does not refer specifically to the plutonium controversy, ICRP No. 26 (1977) states that for late stochastic effects the absorption of a given quantity of radiation energy is likely, in fact, to be less effective when due to a series of 'hot spots' than when it is uniformly distributed. This conclusion was derived after considering both theoretical and epidemiological evidence, and is based largely on the reasoning that high local doses result in the death, or loss of reproductive capacity, of the neighbouring cells. At first this may seem paradoxical, but such highly localized damage appears to be less likely to induce cancer than a less severe, more general, damaging effect. The ICRP report goes as far as to say that to assess the risk by assuming a homogeneous dose distribution would probably overestimate the actual risk for stochastic effects; and that for non-stochastic effects the limited amount of cell loss that might result, at moderate dose levels, would be most unlikely to cause any impairment of organ function.

There is one exception, and that is the skin. Although it is thought that, relative to many other tissues, the skin is less liable to develop fatal cancer after irradiation, it is accepted that cosmetically unacceptable changes in the skin may appear at absorbed doses of 20 Gy (2000 rad) or more delivered over weeks or months to localized areas. It is therefore recommended that routine monitoring of persons occupationally exposed be made on the basis of 100 cm^2 areas of skin.

A selected list of maximum permissible concentrations in air and water, for both a 48- and 168-h week, are given in table 3.13. The levels are those recommended in ICRP 2 (1959) for occupational workers. Although only a selection, the few radionuclides listed in the table are enough to show the many variables which have been taken into account in order to arrive at these recommended concentrations. For isotopes of the noble gases argon, krypton and xenon, only the aerial pathway needs to be considered and there is no critical organ. Similarly there is no critical organ for tritium, but intake both from air and via the gut has to be considered, the former being the more limiting. For the other radionuclides, the chemical state must be considered, if known, because of differences in the behaviour of soluble and non-soluble forms. It will be noted that for a number of radionuclides – ^{60}Co, ^{99}Tc, ^{106}Ru – the lung is one critical organ for insoluble forms and the gut for both forms because these elements are so poorly absorbed. Other radionuclides present a hazard to specific organs when taken into the body in a soluble form; for ^{54}Mn the critical organ is the liver, for ^{59}Fe the spleen, for ^{90}Sr the bone, and for the ^{131}I the thyroid. Other radionuclides, such as those of uranium, have more than one critical organ depending upon the chemical form and route of entry of the radionuclide into the body. There are even factors not associated with the radioactive aspects which must not be overlooked: the maximum permissible concentrations for soluble uranium are based upon a knowledge of its chemical toxicity. Finally, it should again be mentioned that the ICRP data are for exposure over and above natural background; ICRP gives no guidance on the intake of ^{40}K.

It was stated at the beginning of this discussion on internal exposure that, although forming the basis of radiation protection standards throughout the world, the 'critical organ approach' is being replaced. The new system is designed to relate to the methodology adopted in ICRP No. 26. Use is now made of a quantity which allows for the fact that incorporated radionuclides will deliver an absorbed dose over an extended period of time, the absorbed dose being expressed as a dose equivalent. This quantity, the *committed dose equivalent* (H_{50}), is the integrated dose equivalent in a particular tissue which will be received by an individual following an intake of radioactive material into the body. The dose equivalent is integrated over a period of 50 years, considered to correspond to a working lifetime. (If the H_{50} values are multiplied by the appropriate weighting factors (W_T), and summed, a *committed effective dose equivalent* can be derived).

The ICRP therefore recommends that the intake of radionuclides into the body be determined by deriving *annual limits of intake* (ALI) for each radionuclide. Each ALI value will take into account the dose equivalent in each tissue and the relative weighting factors. The concept differs considerably from the previous 'body-burden' approach because, by limiting the intake during each year to less than an ALI, an organ body-burden cannot be achieved until the 50th year of continuous intake at the ALI rate. Recommended ALI values are to be gradually introduced, and will replace such data as are contained in table 3.13.

In summary, therefore, the appropriate quantity to use for comparison with recommended individual dose equivalent limits in the future will normally be the sum of two values: the dose equivalent from external radiation, and the committed effective dose equivalent from internal radiation. The latter will be used to calculate annual limits of intake for different radionuclides. It is important to stress, however, that by using the *effective* value no provision is made for non-fatal somatic diseases, nor for all hereditary effects. Particular circumstances may require that such omissions be taken into account.

It is a remarkable achievement that so much has been learned about the effects on man of one type of hazardous chemicals in so short a time. As a basis of knowledge for evaluating the potential harm to man from their general usage the subject of human radiation protection, or *health physics*, as it is generally termed, has no parallel. There are many classes of chemicals, and indeed hazards of many kinds, which are in every day usage and potentially far more harmful than radionuclides. Nevertheless, as an area of study it will attract a continued and more detailed number of investigations because of the general public concern about the principal future source of radionuclides in the environment – the generation of electricity from nuclear power.

4. Nuclear reactors and their fuel cycles

4.1. *Introduction*

The use of nuclear reactors to generate electricity continues to increase all over the world. By December 1979 about 128 000 million watts (128 GW(e)) were being generated by 249 reactors operating in 22 countries. In common with other principal methods of generating electricity, nuclear reactors produce waste products which require safe disposal. This waste differs from those derived from the burning of fossil fuels in that it is more radioactive, although not all of it is potentially more hazardous. There are, therefore, many aspects of waste disposal which are unique to nuclear power.

Radioactive waste materials arise in three principal ways: from the day-to-day operation of the nuclear reactor – wastes sometimes referred to as 'running releases'; from a number of processes involved in the mining, milling and manufacture of the nuclear fuel itself; and from the subsequent handling and treatment of the spent fuel. The entire cycle actually starts with the mining and milling of the raw material but in order to understand the reasons for different methods of fuel fabrication it is first of all necessary to know something of the nuclear reactor itself and the different types of design currently in operation.

4.2 *Nuclear reactors*

Commercial nuclear reactors are designed to harness the energy liberated in the fission of certain atomic nuclei in order to generate electricity. It was briefly mentioned in chapter 1 that all isotopes of elements of atomic number greater than 83 are unstable. Some very heavy nuclei, such as those of ^{235}U, upon the capture of a neutron to form ^{236}U, will divide into two approximately equal halves; that is undergo fission as discussed in chapter 2. The energy produced in fission is equivalent to the difference in (rest) mass between the interacting particles and the final products. For an atom of ^{235}U the energy released is of the order of 200 MeV. To convert to more every-day units, this is equivalent to about $3 \cdot 2 \times 10^{-11}$ J. Thus the complete fission of 1 g ^{235}U would release $8 \cdot 2 \times 10^{10}$ J, that is $8 \cdot 2 \times 10^{10}$ Ws, $2 \cdot 3 \times 10^4$ kWh; in other words an enormous amount of energy, approximately equivalent to that derived from the combustion of 3 tonnes (t) coal. The power output from a reactor can be desribed either in terms of heat or in terms of electricity produced. For the former the unit megawatt thermal (MWt) is used, and for the latter megawatt electrical (MWe). In fact only about 25 to 35% of the heat output is converted to electrical energy, the rest being dissipated locally as low-temperature heat. The fraction of heat converted usefully would be increased if the temperature at which the reactor operates could be increased, a consideration which has been paramount in modern reactor design.

When a nucleus such as that of ^{235}U undergoes fission, from one to five neutrons may be released. In order for a chain reaction to take place it is necessary for one of these neutrons to induce the fission of another ^{235}U nucleus. On average two to three neutrons are produced and these may escape altogether, take part in non-fission reactions, or induce fission. For ^{235}U the number of neutrons (N) present at any instant is determined by

$$N = N_0 e^{xg} \qquad (4.1)$$

where N_0 is the number of neutrons present at time zero, x is the increase in neutrons per fission and g the generation time, i.e. the number of generations of fissioned nuclei. One mole of ^{235}U (i.e. 235 g) contains 6×10^{23} atoms and thus its complete fission would require this number of neutrons. The fission of 1 kg ^{235}U would therefore require $2 \cdot 6 \times 10^{24}$ neutrons and from equation (4.1) it can be deduced that, with an x value close to unity, this number of neutrons would be obtained after 56 generations. The practical application of such a calculation was quickly appreciated as a means of producing an enormous explosion; the necessary fission generation time would be approximately 10^{-8} s and the energy released equivalent to that of 16 kt of TNT.

Natural uranium consists of three isotopes, ^{235}U occurring at an abundance of 0·7%. The most abundant isotope is ^{235}U. Because of the slight difference in the neutron-to-proton ratios of the two isotopes, the nucleus of ^{238}U requires fast neutrons (>1 MeV) to induce fission whereas fission of the ^{235}U nucleus can be induced by neutrons of any energy, although there is a higher probability of it occurring with slow, thermal, neutrons – in the terminology of nuclear physics, the ^{235}U nucleus has a larger capture cross-section at lower energies. Thermal neutrons, incidentally, are those having that amount of energy attained at thermal equilibrium. For each ^{235}U nucleus which is fissioned by slow neutrons, two to three high-energy neutrons are emitted, but not all of these could necessarily induce fission of either ^{235}U or ^{238}U. They may take part in inelastic collisions with ^{238}U nuclei, which therefore slows them down. Nuclei of ^{238}U may also capture slow neutrons – to form ^{239}U – and thus hinder the continued fissioning of ^{235}U nuclei. In addition, neutrons may be captured by any impurities present and by the products of the fission of other nuclei. This clearly presents a difficulty in attempting to sustain a controlled chain reaction. A further difficulty is the speed with which the fission process takes place. Over 99% of the neutrons produced in fission are released within about 10^{-14} s. These are called the *prompt neutrons*. The remainder, the *delayed neutrons*, are emitted over a period of several minutes; they originate from the excited states of nuclides produced by beta decay, the parents of which are fission products. An example is ^{87}Br which has a half-life of 55s. It decays by emitting a $\dot{\beta}^-$ particle to give ^{87}Kr which, being in a highly excited state, has sufficient energy to emit a neutron immediately to form the stable ^{86}Kr. The occurrence of these delayed neutrons greatly facilitates the control of the fission reaction.

In order to maintain a chain reaction it is only necessary that, as a result of one nucleus undergoing fission, an average of one neutron is produced which induces the fission of another nucleus. This number is called the effective multiplication factor (k). If it equals unity, the number of fissioning nuclei will remain the same; if it is less than unity their number will decrease and the chain reaction will stop; if greater than unity, however, the chain reaction could go out of control. As we have noted, neutrons can be lost and their rate of loss will be affected by the geometry of the system. In an infinitely large core, the ratio of the number of fission neutrons of any one generation to the number in the succeeding generation is signified by k_∞. Thus k, the effective multiplication factor, is actually a product of k_∞ and the probability (P) that fission neutrons will remain in the core of a particular size, i.e. $k = k_\infty P$. Another factor influencing the value of k_∞ is the range of neutron energies present. There are a number of considerations here, but mention may be made of two factors in a reactor consisting of natural uranium and a moderator – such as graphite – which slows the neutrons down. One is the factor p, which is the probability that any fast neutron will be slowed down by the moderator to reach the thermal region – where it can be captured by a nucleus of ^{235}U

-without being captured in non-fission reactions; the second is f, the fraction of thermal neutrons taken up by the uranium nuclei, some of which will undergo fission. It is important that both factors, each less than unity, should be as large as possible. But attempts to increase one factor – such as increasing the relative amount of moderator to increase the value of p – decrease the value of the other. For a homogeneous mixture of natural uranium and graphite, the maximum possible value for k_∞ is 0·8, too low to maintain a chain reaction. There are two ways to overcome this: one is to increase the relative abundance of [235]U present, which would increase the values of both p and f; the other is to alter the geometry, by designing a lattice of large lumps of uranium set in a mass of graphite. This second method increases the value of p but only slightly decreases the value of f, giving an optimum k_∞ value of 1·07. Both methods are used in reactor design.

Before we consider the construction of reactors in general, there is a further important point to be made. The discussion so far has related to the fissioning of [235]U, for which neutrons of any energy can induce fission; such nuclides are described as fissile. The more abundant [238]U, upon the capture of a neutron, becomes [239]U. This radionuclide is short-lived, with a half-life of 23·5 min. It decays by β^- emission to form [239]Np which, by a further β^- emission, with a half-life of 2·35 days, decays to [239]Pu. This last nuclide, like [235]U, is fissile, and several per cent of the total number of nuclei which undergo fission within the reactor are those of [239]Pu. It also follows that the longer fuel is left within the reactor the greater this percentage will be.

Another nuclide, which upon neutron absorption eventually gives rise to a fissile nucleus, is the naturally occurring [232]Th. After capture of a thermal neutron to form [233]Th, a β^- particle is emitted to give [233]Pa which, by further β^- decay, forms [233]U. This last nuclide, like [235]U, is fissile and can sustain a chain reaction. Because both [238]U and [232]Th can give rise to fissile nuclides, they are often referred to as *fertile* nuclides.

In view of the above discussion; what are the basic requirements of a nuclear reactor for the generation of nuclear power? The reactor vessel consists of an active *core* in which the fission chain reaction is sustained; the core contains the reactor fuel – the fissile material – and a moderator if this is required to reduce the energy of the neutrons. The moderator must be constructed from materials of fairly low atomic mass because the nuclei of low-mass elements absorb a large fraction of a neutron's energy and slow it down to the thermal neutron energy region very quickly. The ideal nucleus to remove the maximum fraction of a neutron's energy is one of equal mass – that of hydrogen. A further consideration is that the material should not have a high neutron capture and thus mop up neutrons and prevent a chain reaction from occurring at all. This reduces the choice, from a practical point of view, to water, heavy water (deuterium oxide), carbon in the form of graphite, and beryllium. In some reactor designs the core is also blanketed in a moderating material which acts as a neutron reflector, to minimize the loss of neutrons from the system. The energy of fission is released as heat and reactor operation is dependent upon the ability to remove the heat produced as fast as it is generated. The coolant must therefore circulate through the reactor core so as to maintain, as far as possible, a uniform internal temperature. Because the coolant passes through the core of the reactor it, too, must be a material which will absorb a relatively low number of neutrons. Again both light and heavy water are used and a gas, carbon dioxide, is used in some types of reactor. At high temperatures a liquid metal, sodium, is used. In the majority of reactor designs the coolant transfers heat indirectly, by one mechanism or another, to water to create steam which is used to generate power in a conventional manner by means of a turbine.

Both moderator and coolant are chosen partly on the basis of their poor neutron absorption but, in order to control the reactor, control rods of a *high* neutron absorption material are required – usually boron steel. Several different control rods are used: coarse controls for reactor start-up and shut-down, fine controls for adjustment of the desired operating level, and safety (scram) controls for emergency shut-down. Fission, as we have noted, is virtually an instantaneous process and it would not be possible to regulate it without the presence of the delayed neutrons which, for ^{235}U, represent 0·65% of the total. If this fraction is represented by x, the fraction of prompt neutrons will be $(1 - x)$. Thus of the total number (η) of fast neutrons produced for each thermal neutron observed, $(1 - x)\eta$ are emitted immediately, and $x\eta$ are delayed. Hence the multiplication factor k may again be considered as being affected by two processes; one equal to $k(1 - x)$ and the other equal to kx. When a reactor is started up, the quantity $k(1 - x)$ is adjusted such that it is equal to, or less than, unity. Because x, for ^{235}U, is 0·65% of the total, this can be attained by having k between 1 and 1·0065. The reactor is then said to be in a *prompt critical* state and can become critical by using the prompt neutrons alone. If the power output needs to be raised the control rods are slightly withdrawn and then readjusted. On shut-down the rods are inserted in order to capture the neutrons and reduce the value of k below unity, so that a chain reaction can no longer be maintained.

A reactor consisting of natural uranium and a moderator would not explode, but excess heat could cause the whole system to break apart; a *melt down* as it is called. Each piece would then be sub-critical in size but there would be a high probability of radionuclides being released. The safety rods are therefore typically held up by electromagnets so that they can rapidly be released to shut-down the reactor in an emergency. After a reactor is shut down it cannot be started up again immediately. This is because one of the fission products, ^{135}Xe, has an extremely large cross-section for the capture of slow neutrons. It arises from the decay of a primary fission fragment, ^{135}Te, and has a half-life of 9·1 h. The ^{135}Xe level continues to increase for a while even after the reactor is shut down and it is necessary to wait for it to decay to a sufficiently low level before the reactor can be induced to go critical again.

The reactor core – containing fuel, moderator, coolant and control rods – is surrounded by the reactor shield. This frequently consists of

TABLE 4.1.   *Characteristics of different types of nuclear reactor.*

Reactor type	Typical fuel material	Typical fuel cladding	Moderator	Coolant
GCR (magnox)	Natural uranium as metal	Magnox (Mg alloy)	Graphite	CO_2
LWGR	Uranium dioxide 2% ^{235}U enriched	Zirconium/ niobium alloy	Graphite	Water
CANDU	Uranium dioxide non-enriched	Zircaloy	Heavy water	Heavy water
PWR	Uranium dioxide 2-4% ^{235}U enriched	Zircaloy	Water	Water
BWR	Uranium dioxide 2-3% ^{235}U enriched	Zircaloy	Water	Water
AGR	Uranium dioxide 2-3% ^{235}U enriched	Stainless steel	Graphite	CO_2
HTGR	Uranium carbide 90% ^{235}U enriched	Silicon carbide	Graphite	Helium
FBR	Plutonium dioxide and enriched uranium dioxide	Stainless steel	None	Liquid sodium

an inner thermal shield of iron or steel encased in a biological shield of concrete, the latter being about 2 m thick. The shielding serves to attenuate the gamma rays and neutrons to a level too low to harm the personnel operating the reactor. In some reactor designs the entire reactor is further contained within yet another shield.

Nuclear power reactors have been developed independently by a number of countries and they differ in detail to a varying degree. The basic components of different reactor designs are summarized in table 4.1. and their current usage around the world listed in table 4.2. One of the first types of nuclear reactor to be used for the production of electricity on a commercial scale is close to the basic type of reactor discussed above. It was developed principally in the United Kingdom and France, and is generally referred to as the *magnox* reactor because the fuel, in the United Kingdom, consists of natural uranium metal clad with a thin layer of a magnesium alloy. It is a type of gas-cooled

TABLE 4.2.   *Major nuclear reactors in operation, or in power ascension phase, 1 March 1977.*

Country	GCR type	LWGR	CANDU type	PWR	BWR	AGR	HTGR	FBR	HWGCR
Argentina			1						
Belgium				4					
Bulgaria				2					
Canada			8						
Switzerland				2	1				
Czechoslovakia									1
German DR				3					
Germany FR			1	5	6		1		
Spain	1			1	1				
France	7			1				1	
Great Britain	26					5		2	
India			1		2				
Italy	1			1	1				
Japan	1			7	7				
Netherlands				1	1				
Pakistan			1						
Sweden				1	4				
USSR		12		7	5			2	
USA				35	25		1		

From IAEA *Power Reactors in Member States*, 1977 edition; courtesy of the International Atomic Energy Agency, Vienna.

reactor (GCR). The *fuel elements*, which typically number about 26 000 for a reactor core producing 300 megawatts of electricity (300 MW(e)), are stacked in channels in a massive pile of graphite blocks. The fuel elements contain 320t natural uranium. The coolant consists of carbon dioxide gas which, under high pressure, is forced through the fuel element channels leaving, in the later designs, at a temperature of about 360° C. The gas passes through a heat exchanger and is returned to the reactor. Reserves of carbon dioxide are held in case of an emergency resulting in loss of coolant; although, providing the control rods can be inserted quickly – there are sufficient to allow for failure here too – enough gas can be kept flowing at atmospheric pressure to cool the fuel.

A simplified plan of a magnox reactor is shown in figure 4.1. In the earliest types, the reactor pressure vessel was contained within a steel sphere connected by ducts to the steam-generator units. Later designs

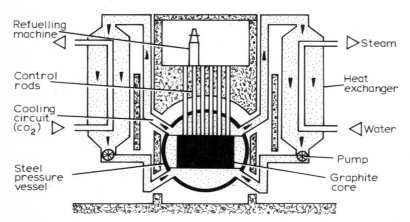

Figure 4.1.   Schematic diagram of a magnox reactor.

consist of pre-stressed concrete pressure vessels with an integral arrangement of steam generators; this allowed more than a two-fold increase in reactor size. The reactor core of a magnox reactor is large relative to those of other designs and requires a more or less continuous refuelling process. For this to be done while the reactor is on-load requires a complex fuel-handling system which, operated by remote control, removes spent fuel rods and inserts new ones. A variation of the graphite-moderated reactor, which does not use natural uranium, has been used in the USSR. These reactors have ordinary light water instead of carbon dioxide as coolant; they are therefore referred to as light water graphite reactors (LWGR).

The alternative major reactor design to use natural uranium – as uranium dioxide pellets – is that developed in Canada. These reactors are know as CANDU reactors (Canadian deuterium uranium), or more generally as PHWRs (pressurized heavy water reactors). The reactor core (figure 4.2), instead of consisting of a graphite block, is

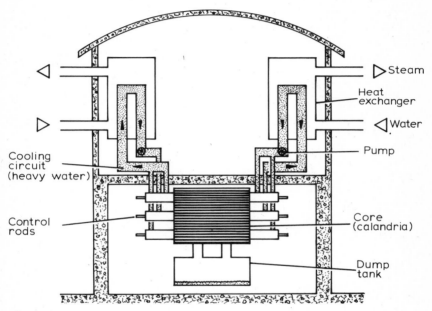

Figure 4.2.   Schematic diagram of a CANDU reactor.

made of a cylindrical stainless steel tank, called a *calandria*, through which run a number of horizontal pressure tubes made of zircaloy, an alloy of zirconium. The calandria is filled with heavy water – deuterium oxide – which acts as a moderator, and the coolant is also composed of heavy water, under very high pressure to prevent it from boiling. It leaves the reactor at a temperature of ~300° C. The control rods are situated in the side of the calandria and a 'dump' tank beneath enables the entire volume of the moderator to be removed from around the pressure tubes in the event of an emergency. Again because of its large size necessitated by the use of natural uranium, a complex 'on-load' refuelling process has had to be developed. There are two refuelling machines, on opposite faces of the reactor. One machine inserts bundles of new fuel rods into the pressure tubes and the other collects the spent fuel rods as they emerge on the other side.

In the USA the development of commercial power reactors arose from the design of reactors required to power submarines. Nuclear power is an ideal form of power for submarines because, in contrast to the burning of fossil fuels, it does not require oxygen. A compact design was derived (figure 4.3.) consisting of a single fuel element

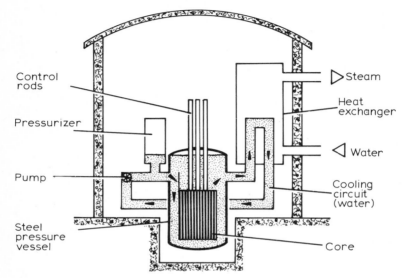

Figure 4.3.   Schematic diagram of a light-water (PWR) reactor.

assembly of up to 200 zircaloy fuel 'pins', each 3·5 m long, immersed in a large steel pressure vessel containing ordinary 'light' water. They are therefore known as pressurized water reactors, or PWRs. The light water serves both as moderator and coolant, but because it has a higher neutron-absorbing capacity than heavy water, it is necessary to increase the percentage of ^{235}U in the core. The fuel therefore consists of uranium dioxide in which the fraction of ^{235}U has been enriched from 0·7% to between 2 and 4%. The pressure vessel contains the reactor core, control rods which pass through the lid, and the light water under pressure. Typical operating pressures are about 13·8 to 17·2 MPa (2000 to 2500 psi) so that the water attains a temperature of around 270° C without boiling. The water passes in a closed circuit to a heat exchanger, the circuit including a *pressurizer* to maintain the pressure by either heating or cooling an appropriate quantity of the water. In order to refuel the reactor it is necessary to shut it down completely, remove the lid, and replace an appropriate portion (about-half) of the fuel pin assembly. This takes place at intervals of 12 to 18 months. An obvious potential danger in this type of reactor is the possible rupture of the cooling system tubing, there being no inherent way of preventing the reactor from overheating if the coolant was suddenly lost. Their design therefore incorporates a number of emergency core-cooling systems (ECCS) and the entire reactor is housed within a pressure containment building, usually double-walled.

If water is allowed to boil it is more efficient at removing heat. One form of light water reactor has therefore been developed such that the coolant/moderator within the reactor core is allowed to boil, the steam generated ($\sim 280°$ C) being separated, dried, and passed direct to the turbine generators. Having gone through the generator, the steam is condensed and passed back into the reactor core. The general lay-out of such a boiling water reactor (BWR) is shown in figure 4.4.

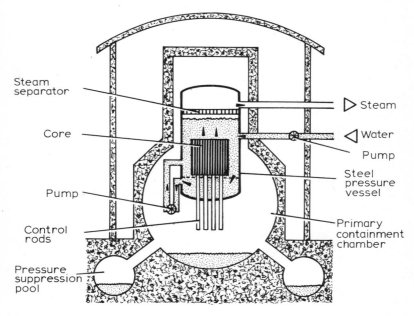

Figure 4.4.   Schematic diagram of a boiling water (BWR) reactor.

The fuel is ^{235}U, enriched, as uranium dioxide. The core assembly generally resembles that of a PWR except that, because the steam-collecting assembly is on top of the reactor, the control rods are inserted from below; they cannot therefore be lowered by gravitational force in an emergency. Because the reactor is linked directly with the generator, provision must be made for the containment and re-routing of the steam in case of generator malfunction. The entire reactor is housed within a primary containment chamber which incorporates, underneath, a large ring-shaped tunnel partly filled with water. Any escaping water or steam enters this tunnel – the pressure suppression pool – and condenses. There are, in addition, emergency core-cooling systems. As with the PWRs, the reactor has to be shut down for refuelling.

All of the types of reactor described above are currently in operation. They have a number of drawbacks as producers of electricity. Both PWRs and BWRs have a rather low efficiency for the conversion of heat to electricity because of the relatively low temperatures produced. The magnox reactors, although attaining higher temperatures, are limited by the type of fuel used. Uranium metal has been found to suffer a number of complex structural changes within the reactor core which place limitations on its burn-up time. In addition, magnox ignites at 645° C, even in an atmosphere of carbon dioxide. Uranium metal will also form the oxide at high temperatures, and this is an exothermic reaction. In spite of the relatively higher temperatures which could be safely attained, the efficiency of magnox reactors is not greater than that of the PWRs and BWRs. In the United Kingdom the magnox reactors are now being succeeded by a generation of advanced gas-cooled reactors (AGRs). These are designed to achieve higher gas temperatures and greater burn-up of the fuel. As with the USA reactor designs, the fuel now used is in the form of uranium dioxide which, although more suitable for higher temperatures, has a lower thermal conductivity. Uranium dioxide fuel rods are therefore much thinner than those used in magnox reactors. For AGRs these are encased in stainless steel rather than zircaloy. Stainless steel is a reasonable absorber of neutrons and thus the fuel has to be enriched to 2 to 3% ^{235}U. Graphite is still used as a moderator, encased in a prestressed concrete pressure vessel, and the coolant gas emerges at a temperature of ~640° C. As with the magnox reactors, the core is refuelled on-load.

Both the United Kingdom and Canada have sought to combine the best features of the CANDU reactor and the direct cycle BWR. Experimental reactors have been built using heavy water as moderator and light water as coolant. Heavy water gas-cooled reactors (HWGCRs) have been built in Czechoslovakia and France. Other designs have been developed in the United Kingdom, the USA and West Germany. The principal limitations of the reactors described above again relate to the temperatures which can be attained. In order to achieve increasingly higher temperatures, reactor cores have been designed on an entirely different basis, using ceramic materials instead of metals for fuel. These reactors, also gas-cooled and thus referred to as HTGRs (high temperature gas-cooled reactors) have a completely different arrangement of fuel and moderator. The fuel of the two commercial reactors presently operating consists of a mixture of uranium dicarbide and thorium dicarbide particles. Each particle is coated with a layer of pyrolytic carbon to minimize the escape of fission products. The fuel is contained within graphite and the coolant gas used is helium which, unlike carbon dioxide, will not react with the graphite at very high temperatures. Helium leaves the reactor core at a temperature of ~800° C in the USA design and ~950° C in the West German design. The latter reactor uses a pebble-bed concept in which the helium is blown upwards through the fuel, embedded in graphite spheres of about 2·5 cm diameter, contained in a large bin.

The sustained fission chain reaction in all of these reactors is based on ^{235}U. This is the only naturally occurring fissile nuclide and, as we have noted, constitutes only 0·7% of the natural uranium; it places a severe limitation on the utilization of nuclear power unless the fertile properties of ^{238}U, and ^{232}Th, are exploited. In fact, even in the use of ^{235}U fission, some of the created ^{239}Pu also undergoes fission and contributes to the heat output; but the amount of ^{239}Pu produced is less than the amount of ^{235}U used. It is possible, however, to construct a reactor which creates more fissile material than it uses: such reactors are called *breeders* – they are of an entirely different design. To breed new fissile nuclei in a chain reaction requires that as a result of the fission of one nucleus, on average, one neutron continues the chain reaction and one is captured by a fertile nucleus such that it creates a further fissile nucleus. In practice some neutrons take part in other reactions, or are lost, so that more than two neutrons per fission are required. There are many technical difficulties in attaining these conditions. Fissions caused by fast neutrons produce, on average, more new fast neutrons than those caused by thermal neutrons. But it takes many more fast neutrons to cause a fission; and fast neutrons are required to convert ^{238}U to ^{239}Pu. One difficulty in reactor design is that a much higher neutron density must be created and thus the core has virtually no moderator, a minimum of structural materials, and as little coolant as possible. Also, because of differences in the chances of neutron capture at different neutron energies, it is found that to increase the production of ^{239}Pu to a point where it is greater than the loss of the original material it is better to use ^{239}Pu as fuel rather than enriched ^{235}U – or at least a mixture of the two. The core of a fast breeder reactor (FBR) therefore consists of a very compact arrangement of fuel assemblies – plutonium and uranium oxides clad in stainless steel – surrounded by a blanket of ^{238}U derived from 'depleted' uranium, that which has had ^{235}U removed for enrichment purposes. The coolant used is sodium, a metal which melts at a temperature of 97·5° C. With a boiling point of 883° C it does not need to be pressurized, which greatly facilitates engineering design. Other advantages are that it has a high thermal conductivity, and does not readily absorb fast neutrons. It does, nevertheless, absorb some neutrons to form ^{24}Na, a strong gamma emitter. The cooling system therefore has to be contained within the biological shield and the heat exchanged to a secondary sodium cooling circuit. The secondary sodium circuit passes through thin tubes to create steam and it is here that some difficulties were originally encountered: the seals must be very reliable because sodium reacts violently with water. A simplified diagram of a fast breeder reactor is shown in figure 4.5.

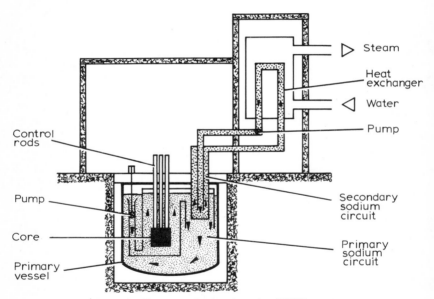

Figure 4.5.   Schematic diagram of a fast breeder (FBR) reactor.

Brief mention may be made of fast breeder operation. At higher neutron speeds there are fewer delayed neutrons: this is a disadvantage, but one advantage is that the reactivity of the reactor – the quantitative extent to which the reactor tends to be critical – decreases as the temperature rises. This is due to an effect known as Doppler broadening which results in an increase in the range of neutron energies which different species of nuclei can absorb. Fortunately, at higher temperatures, there is a greater chance that neutrons will be absorbed by ^{238}U nuclei than by ^{235}U or ^{239}Pu nuclei. This helps to moderate the tendency, which would otherwise obtain, for the reactor to become more difficult to control as its temperature increased.

Prototype breeder reactors are already in use in the United Kingdom, France, West Germany and the Soviet Union, all based on the production of ^{239}Pu from ^{238}U. The ^{239}Pu could either be used in further breeder reactors or used in thermal reactors in place of ^{235}U; it could also be used to make nuclear weapons, and thus breeder reactor development is the subject of much political debate. An alternative breeding cycle, based on the production of ^{235}U from ^{232}Th is also being considered, but at present no such commercial reactor exists.

One question always raised is: are reactors safe? This, of course, is a simplistic question to which there is no answer. The principal areas of concern centre around the possible consequences of not being able to shut down the reactor in a potentially dangerous situation; and whether the residual heat arising from the fission product decay could be removed fast enough from the core. Failure to meet these possibilities would result in pressure developing sufficient to rupture the containment buildings and thus release large quantities of radionuclides into the environment. All reactors contain a complex array of instrumentation which measures neutron densities, temperature, pressure and so on; they also contain a number of safety arrangements designed to be as fail-safe as possible. Throughout fabrication and construction of any nuclear plant, great attention is paid to what may be described as quality control; and all reactors must meet specific requirements laid down by licensing authorities. Nevertheless it cannot be stated that an accident will not happen, as the incident in 1979 at Three Mile Island (USA) has shown, only that the chance of its happening has a very low probability. Giving numerical values to such probabilities is only useful in as much as they compare favourably with the probability of other improbable occurrences – being struck by lightning or by meteors, or being in a domino effect as the result of the collapse of the Empire State Building in New York city. What can be calculated, however, are *maximum credible accidents* and such calculations have usually been required as a basic reactor licensing criterion. Even these calculations do not allow, by definition, for the incredible happening. But in fact numerical values derived from various calculations, although of some concern, are not usually as disastrous as one might imagine. Studies in the United Kingdom, a fairly densely populated country, have calculated that as a result of containment rupture of a 1000 MWe reactor in which 10% of gaseous and volatile products escaped, the release of some 370 PBq (10 MCi) of ^{131}I and 37 PBq (1 MCi) of ^{137}Cs would be of the greatest environmental concern. If this occurred in a coastal area where the surrounding population, varying in density, numbered 420 000 within a 24km radius, the following possible outcome has been calculated.† There would be a 30% probability that between 100 and 1000 people would develop thyroid cancers over a period of 10 to 20 years, and a 20% probability that this number would lie between 1 000 and 10 000. A small fraction of these would prove fatal. Other cancers would also occur but the probable number of fatalities is estimated as being 100 to 150 deaths from thyroid cancer and 10 to 200 deaths from leukaemia and lung cancer – about the same number of people carried by a jumbo jet. These figures may well vary by an order of magnitude either way, depending upon conditions at the time of the accident, the most notable effect being the direction and strength of the wind. Wide-scale contamination would cause a very difficult and expensive clean-up operation and undoubtedly result in severe inconvenience to the majority of the surrounding population. Such calculations as these do not necessarily allay public anxiety; indeed the mere probability, no matter how improbable, is sufficient to initiate it. Attention to details of safety – particularly to new reactor designs such as the FBRs – therefore continues to be one of the areas of nuclear power production to which continued research and development is made.

Even very minor reactor accidents are not, in fact, the means by which radioactivity enters the environment from nuclear reactor operation. From the descriptions of different reactor types given above it would appear that, in normal operation, the reactor core and its fuel is completely enclosed and sealed from the outside world: this is an oversimplification. Both gaseous and liquid radioactive wastes arise, and some of these are deliberately released into the local environment. Gaseous releases originate in a number of ways. Some forms of magnox reactors use a thin layer of air drawn from the outside to cool the inside wall of the concrete pressure vessel. The air stream, although not coming into contact with the reactor core, is nevertheless subjected to neutron bombardment. Air contains about 1% of the inert gas argon and the most important neutron activation product thus formed is ^{41}Ar, a nuclide with the short half-life of 1·8 h; approximately 740 GBq h^{-1} (20 Ci h^{-1}) are produced from this source. Any particulate matter carried through will also become irradiated and the cooling air is therefore passed through glass-fibre filters before release back to the environment. In addition, some carbon dioxide coolant inevitably leaks from gas-cooled reactors through

† Royal Commission on Environmental Pollution, 1976, Sixth Report Cmnd. 6618. Chairman Sir Brian Flowers (London: HMSO).

valve seals and other permeable joints. All of the areas which handle coolant gas, and the main enclosures of the reactor pressure vessel, are swept by a ventilation flow of air. Before this air is discharged to the atmosphere it is passed through high-efficiency .particulate filters (HEPAs). The direct neutron activation of ^{16}O in the gas gives rise to ^{16}N, and some ^{35}S is also formed from ^{34}S and ^{35}Cl impurities. Variable amounts of ^{14}C are also produced, both from ^{13}C in the graphite moderator and from ^{14}N, another impurity. Every time a refuelling machine is connected to the reactor the machine itself acquires some of the coolant gas, and this is discharged to the atmosphere. A fraction of the coolant gas is also routinely bled off and replaced. These discharges, which may have a variety of particulate matter in them, are passed, before release, through filters which typically remove particles down to $5 \mu m$ with an efficiency of 100%, and particles down to 2 μm with an efficiency of 95%. In PWRs, ^{14}C is produced from ^{17}O in the oxide fuel and in the moderator, and from ^{14}N present as impurities. Some ^{14}C also arises from ternary fissions. Ternary fissions also give rise ^{3}H in all reactor types; in magnox reactors it also arises from neutron activation of lithium impurities in the reactor core graphite. This is removed as tritiated water when the coolant gas is passed through a special drying plant designed to maintain its condition. The tritium thus joins the liquid effluent arising from the plant.

Of greater importance are leakages from the fuel elements into the reactor coolant. The fuel elements remain within the reactor core for up to about 3 years. During this period fission products are formed within the elements themselves, and their claddings – although made of materials with a low neutron absorption capacity – also gradually accumulate neutron activation products. The cladding is extremely thin and is subjected to thermal and mechanical stresses and corrosive action by the coolant. The build-up of gaseous fission products, particularly nuclides of xenon and krypton, generates internal pressure. Nuclides of helium and hydrogen are also formed by a variety of reactions and these may diffuse through some forms of cladding, or at least cause embrittlement. It is inevitable, therefore, that small cracks may develop in some fuel elements, allowing the slow release of some of their contents into the coolant. The short-lived ^{131}I is important here, particularly in gas-cooled reactors. Corrosion of the entire coolant circuit gradually results in a build-up of neutron activation product nuclides, such as ^{59}Fe, ^{51}Cr, ^{54}Mn and ^{60}Co. Should a fuel element burst, a fairly large amount of radioactivity would be released. In those reactors which have on-load refuelling, such as the gas-cooled reactors, a complex array of detection equipment very quickly records that such an event has taken place; the fuel channel is then located and the offending fuel can removed and replaced. In the water-cooled reactors such action would require complete reactor shut-down. They are therefore designed to operate with up to 1% fuel element failure, and provision is made for the continual diversion of part of the primary coolant. The diverted water is passed through anion and cation exchange resins, and possibly also through gas-stripping devices, before being returned to the primary coolant main-loop. This process also allows other procedures, such as pH adjustment, to be made. Boiling water reactors, in which the primary coolant, in the form of steam, is passed direct to the generators, will clearly give rise to a greater amount of radioactivity to be disposed of locally because of leakage in the gland-seals of the turbines, condenser vacuum pumps and so forth. Such gases pass to delay tanks and may also pass over charcoal bed filters.

There are other sources of both liquid and solid waste. Liquid wastes arise from drainage, laundry effluent, washing facilities, etc., and a miscellany of solid wastes such as filter bed materials and contaminated benchware also continually builds up. All of these, including the by-products of contaminated coolants discussed above, are disposed of, under authorization, in a variety of ways. One important waste has not been mentioned, however, and that is the radioactive waste arising from the fuel element after it has been removed from the reactor. The fuel element contains virtually all of the radioactivity arising from nuclear power and before discussing, in chapter 5, the means by which radionuclides are introduced into the environment in a controlled manner as a waste management policy, it is necessary to examine the processes involved in making, using and disposing of the fuel element itself; processes referred to collectively as the nuclear fuel cycle.

4.3.   *The nuclear fuel cycle*

Uranium ores occur widely in many countries and in a variety of complex forms such as uraninite, uranothorite, thorianite and carnotite. Major reserves are those in the USA – notably in the area of the Colorado plateau spanning the states of Utah, Colorado, Arizona and New Mexico, and also in the state of Wyoming; other major reserves occur in Canada, Australia, South Africa, Namibia, Niger, France and Zaire. Uranium-bearing minerals often occur in mines in conjunction with other ores, as in the silver mines of Czechoslovakia, the gold mines of South Africa and the cobalt and silver mines in Canada's Northwest Territory. A few mines, such as those in Zaire, are opencast, but the majority are hardrock underground mines. The principal hazard to the miners is the ^{222}Rn and its daughters already discussed in chaper 2. The concentration of radon daughters is expressed in a unit called the working level (WL), not a very apt term. The WL is defined as *any* combination of radon daughters which, in 1 litre air, will result in the ultimate emission of $1 \cdot 3 \times 10^5$ MeV potential alpha energy; it corresponds to an activity concentration of $3 \cdot 7$ Bq l^{-1}(100 pCi l^{-1}) ^{222}Rn in equilibrium with its short-lived daughters. There is no doubt that miners of uranium have, in the past, died from cancers of the respiratory system as a result of inhaling radionuclides in poorly ventilated mines. Inevitably, strict correlations between exposure and the incidence of such cancers have been difficult, particularly when habits such as smoking have also to be considered.

Radon daughters may attain secular equilibrium with radon in static conditions. This situation is minimized as far as possible in mines by maximizing the rate of ventilation. This reduces the concentration of the daughters, which associate with dust particles, more than that of the radon diffusing into the air from water and rock surfaces. The degree of exposure to mine workers thus depends upon the degree of ventilation. The working environment can be monitored in a number of ways and the only people at risk are the miners themselves; there is virtually no general environmental hazard.

Uranium ore usually needs to be at least $0 \cdot 05\%$ by weight of U_3O_8 if it is to be mined economically at present. The uranium is extracted and begins its cycle (figure 4.6) by being concentrated in a process referred to as *milling*. The ore is crushed and ground and the uranium leached out by a variety of methods, usually dependent upon the lime content of the ores. There then follow a variable number of complex treatments involving some form of decantation, filtration, flocculation and so on, culminating in a recovery process achieved by chemical precipitation, ion exchange or solvent extraction. These processes result in a dried powder containing 70 to 90% by weight of uranium as U_3O_8, or its equivalent. When it is in the form of ammonium diuranate it is referred to as 'yellow-cake'. The milling process thus separates the uranium isotopes from the original ore, the majority of the uranium daughter products remaining in the depleted ore – the *tailings*. Tailings are usually pumped into ponds where solids can settle out and shorter-lived radionuclides decay away. The resultant clear liquid may be filtered, neutralized, and discharged into streams or allowed to evaporate off. The final waste product inevitably contains a large

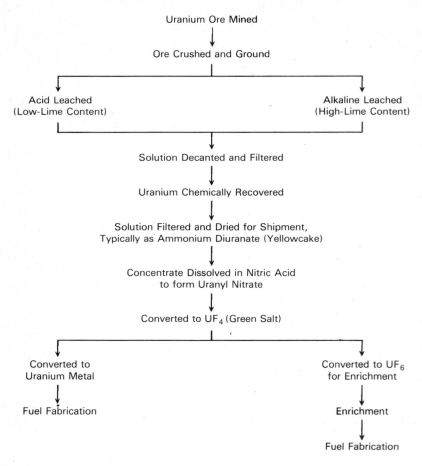

Figure 4.6.   Stages in the treatment of uranium for fuel fabrication.

quantity of radium and its daughters, their rate of emanation being dependent upon the exposed surface area and weather conditions – particularly wind speeds. There may be further complications when the tailings are extremely fine and prone to being wind blown.

The disposal of tailings has had a somewhat chequered history. The direct contamination of river water in southwestern Colorado, USA, at one stage resulted in the ICRP-recommended maximum permissible concentration of [226]Ra in drinking water being exceeded by a factor of three. Elsewhere in Colorado tailings have been used for road material and land fill, the most notorious example being that at Grand Junction, where not only was it used as land fill on which buildings were constructed, but it was even used in some construction materials. This has resulted in radon levels within some of these buildings exceeding WL limits set for miners. Because of these and other gross oversights much greater thought, planning, and tighter legislation have resulted. Tailings are now stabilized with topsoil, concrete sheets, asphalt and so on, and consideration has even been given to their storage in dry mines.

Uranium for reactor fuel must be especially free from elements having a high neutron-absorbing capacity and thus the uranium concentrates are chemically purified, usually following dissolution in nitric acid. The uranyl nitrate thus formed is then converted to uranium tetrafluoride, known commercially as 'green salt.' Subsequent treatment depends upon the type of fuel to be manufactured. For magnox reactors, which use natural uranium, the green salt is reduced by magnesium and the uranium remelted under vacuum, alloyed, and cast into rods. The rods are heat-treated and machined to size. Fuels which are not based on natural uranium need to be enriched. It will be recalled that in chapter 1 it was stressed that the difference between one isotope and another of the same element lay in the properties of the nucleus, not in the number of orbiting electrons. It therefore follows that one cannot separate different isotopes of the same metal by a selective chemical process because the isotopes would behave, chemically, exactly the same. For such a heavy element as uranium, however, there is a slight but significant difference in the mass of ^{235}U and ^{238}U, a difference of little over 1%; it is not very great but sufficient to allow separation of the two by physical methods. The most commonly used process to achieve such a separation has been based on the effect this slight difference in mass has on the ability of the two isotopes to diffuse, in the gaseous state, through a thin porous membrane. The lighter ^{235}U atoms diffuse slightly faster through the membrane than those of ^{238}U. The most convenient uranium compound with which to effect this enrichment process is uranium hexafluoride (UF_6), commonly known as *hex*, a compound which sublimes at temperatures above 56·4° C. It is a highly corrosive, reactive gas and therefore needs very careful handling. The hex, under controlled pressure, enters a metal cell divided by a thin porous metal membrane. Gas from the lower-pressure side of the membrane is subsequently slightly enriched in ^{235}U relative to that of the gas on the higher-pressure side. Even 'slightly' is a rather strong word for, in fact, the low-pressure gas is about 1·0043 times as rich in ^{235}U. The process thus has to be repeated many times in series, the low-pressure gases being passed along in cascade and the high-pressure gases being returned to lower stages. The process becomes successively easier and the greatest effort is required in the earlier stages. For fuel fabrication for all but the high-temperature reactors, the process is repeated until the hex is enriched from the natural 0·7% ^{235}U content up to 2 to 4% ^{235}U content. The 'waste' produced is the depleted uranium which contains between 0·25 and 0·3% ^{235}U, a fraction which is too low to make continued extraction worthwhile.

The entire gaseous diffusion method for uranium enrichment is extremely energy-intensive. It requires an enormous complex of pumps, condensers and such like, and because the physical process of pumping hex around increases its temperature, large-scale cooling of the plants is required. The process is so energy-intensive – it is estimated, for example, that 2·5% of the electricity generated by an AGR reactor would be required to enrich uranium for its fuel – that other means of enrichment have been sought. One method recently employed in the United Kingdom is a process which replaces a diffuser with a centrifuge. The gas, again hex, enters a spinning centrifuge. The heavier ^{238}UF$_6$ molecules drift by centrifugal force towards the outer perimeter and the gas remaining in the centre is thus slightly enriched in ^{235}U. Once again the degree of enrichment is very slight and the process has to be continued in cascade many times before the required degree of enrichment is attained. The method is considerably less energy-intensive than the gaseous diffusion process. Other enrichment processes have also been mooted, the most elegant being that in which a highly tuned laser is able to selectively ionize ^{235}U in a natural uranium hex gaseous mixture.

Enriched uranium is converted to uranium dioxide for use in PWRs and BWRs, and natural uranium is converted to uranium dioxide for CANDU reactors. A variety of techniques are used but all attempt to create pellets of uranium dioxide which are as dense as possible. The entire process requires a high degree of skill, involving baking at high temperature, sintering (compression under heat) in hydrogen, washing and grinding. The size and shape of the finished product has a great influence on the properties of the fuel – thermal conductivity for example – and thus each process involves careful inspection.

The fuel, however produced, is clad in a suitable material which, as we have seen, differs from one type of reactor to another. The cladding, apart from facilitating insertion and removal of the fuel from the core, must also be able to withstand high temperatures, retain fission products, protect the fuel from the coolant, have a high thermal conductivity, and a neutron absorption capacity as low as possible. For the magnox reactors the material is a magnesium-based alloy, for AGRs stainless steel, and for PWRs, BWRs, and CANDU reactors the zirconium-based alloys. The fuel is loaded into tubes, and the tubes filled with helium, capped, and arranged into fuel-element assemblies suitable for the reactor. Each stage is very carefully inspected, for once in the reactor its possible malfunction is a major inconvenience.

The manufacture of fuel elements, as described, presents few problems from an environmental point of view. One of the principal dangers of uranium is its chemical toxicity, which is similar to that of lead; working conditions are therefore closely monitored. The production of enriched uranium has obviously to be undertaken carefully and precautions taken to avoid it being stored in sufficient bulk for it to attain criticality. Great care must also be taken to prevent water entering any storage area because this would act as a moderator and increase the danger. The manufacture of plutonium-based fuel elements is altogether a more difficult process. As a metal it has a number of different crystal phases, it has a low melting point and oxidizes violently on contact with air. Reactor-grade plutonium also consists largely of fissile nuclei and thus even greater care is necessary to avoid an unsuitable juxtaposition which, if resulting in criticality, would result in considerable contamination of the site.

The transport of fuel to the reactors is subject to the statutory rules and regulations which apply to the transport of all radioactive substances. These rules and regulations differ somewhat from country to country and are inevitably complex because of the variety of radioactive substances which are now in everyday use; all are based on IAEA recommendations. In the USA, for example, there are seven 'transport groups' depending upon the relative potential hazard of the nuclides involved. For nuclear fuel on its way to a reactor the quantity of radioactivity to be considered is relatively low, but great caution is obviously necessary to ensure that enriched uranium fuels are not placed in a critical combination. In fact the greatest concern is that of mechanical damage and the elements are crated in specially designed packages, sometimes referred to as 'birdcages'. At the reactor the elements are checked and stored for use; they are again carefully inspected before entering the refuelling machine. Each element bears a unique reference number and a stock control system accounts for each one: its arrival, position in the reactor, removal from the reactor and exit from the site.

Providing all goes well, the fuel element remains within the reactor core for 2 to 3 years. As we have noted, some types of reactor have a continual refuelling system whilst others do not. There are many factors which limit the time spent by the fuel within the reactor, including the absorption of neutrons by the fission products – which decreases the multiplication factor – the physical strains of increased internal pressure resulting from fission product formation, and external corrosive effects caused by the circulating coolant. For those reactors which are refuelled in a single operation this would imply that the reactivity of the reactor would continually decrease, making it difficult to maintain a constant heat output. This can be offset somewhat by initially introducing a neutron absorber, such as boron (^{10}B), into the fuel, its cladding or to the coolant. The absorber, or *burnable poison* as it is usually called, gradually transmutes, as a result of neutron absorption, to other isotopes which have a lower neutron absorbing capacity. Thus the gradual loss of this neutron absorber helps to compensate for the loss of reactivity resulting from fuel burn-up.

Eventually, however, the fuel element must be removed from the reactor. The time planned for a fuel element to spend within the reactor core varies from one type of reactor to another. For a magnox-type reactor a typical dwell-time is about 1000 days; for the AGR it is about 1800 days and for PWRs and BWRs it is typically 1200 days. An increase in the amount of power obtained per tonne of uranium (MWe tU^{-1}) has been one of the principal aims of modern reactor design. For example, the mean fuel rating of the magnox reactors is about 3 MWe tU^{-1} and that of the AGR has been improved to about 10 MWe tU^{-1}; thus for the duration of the fuel in the reactor one obtains 3000 and 18 000 megawatt days per tonne of uranium, respectively.

Upon removal from the reactor, the fuel element will still contain a large fraction of ^{235}U which has not undergone fission, and some ^{239}Pu which has been created but has not undergone fission. The fuel element is now very radioactive and very hot. The majority of its radioactive content is that of fission products, retained by the cladding, but the cladding itself will contain radionuclides formed by neutron absorption – the neutron activation products. The spent fuel elements are placed into a heavily shielded chute and transferred, in the majority of cases, to a specially constructed *cooling pond* alongside the reactor; a few reactors store spent fuel rods in gas-cooled chambers. There are initially a very large number of radionuclides present in the fuel element but many of these have very short half-lives. As they decay the heat generated also declines. By the end of some three months the number of radionuclides present in any quantity will therefore have decreased considerably (table 4.3). In fact there are few radionuclides with half-lives greater than 2 months.

TABLE 4.3. *Approximate theoretical quantities of some major fission products, in TBq (kCi), at different times after removal from a reactor that has operated at 1MW(e) for 1 year.*

Radionuclide	At $t = 0$		At $t = 100$days		At $t = 5$years
^{85}Kr	7·07	(0·191)	6·92	(0·187)	4·88 (0·132)
^{89}Sr	1413·40	(38·2)	381·10	(10·3)	
^{90}Sr	52·91	(1·43)	52·54	(1·42)	44·40 (1·2)
^{90}Y†	52·91	(1·43)	52·54	(1·42)	44·40 (1·2)
^{91}Y	1809·30	(48·9)	536·50	(14·5)	
^{95}Zr	1820·40	(49·2)	629·00	(17·0)	
^{95}Nb†	1783·40	(48·2)	1061·90	(28·7)	
^{103}Ru	1143·30	(30·9)	219·04	(5·92)	
^{103m}Rh†	1143·30	(30·9)	219·04	(5·92)	
^{106}Ru	80·66	(2·18)	66·60	(1·8)	2·59 (0·07)
^{106}Rh†	80·66	(2·18)	66·60	(1·8)	2·59 (0·07)
^{131}I	932·40	(25·2)	0·15	(0·004)	
^{133}Xe	2046·10	(55·3)			
^{137}Cs	39·96	(1·08)	39·59	(1·07)	35·89 (0·97)
^{137m}Ba†	38·11	(1·03)	37·74	(1·02)	34·04 (0·92)
^{140}Ba	1912·90	(51·7)	8·51	(0·23)	
^{140}La†	1912·90	(51·7)	9·99	(0·27)	
^{141}Ce	1768·60	(47·8)	175·38	(4·74)	
^{143}Pr	1676·10	(45·3)	10·73	(0·29)	
^{144}Ce	987·90	(26·7)	769·60	(20·8)	9·99 (0·27)
^{144}Pr†	987·90	(26·7)	769·60	(20·08)	9·99 (0·27)
^{147}Nd	806·60	(21·8)	1·48	(0·04)	
^{147}Pm†	181·30	(4·9)	177·60	(4·8)	50·32 (1·36)

†Daughters.

From IAEA Technical Report Series No. 152, 1974; courtesy of the International Atomic Energy Agency, Vienna.

The cooling pond is filled with water which acts both as a coolant and as a biological shield. A typical depth is 6 m. The fuel elements are held in skips which can be moved, if necessary, around the ponds.

There are several possible means by which radioactivity may appear in the pond water: from corrosion of the cladding material – releasing neutron activation products; from internal leakage through corrosion-induced faults in the cladding – releasing fission products; and from the solution of deposits formed on the outside of the fuel cladding, deposits arising from the coolant's content of corrosion products from anywhere within the reactor's primary cooling system. Corrosion can be minimized by careful control of water quality. Magnox cladding is fairly rapidly attacked by water, and conditions are held such that an oxide or hydroxide film rapidly forms over them. This is achieved by maintaining a high (11 to 11·5) pH. Corrosion is further minimized by using demineralized water to reduce the concentration of chloride, sulphate, and carbonate. The pond cooling water system is designed such that its entire volume passes, each day, through heat exchangers; it also passes through filters and ion-exchange resins. The filters and resins not only help to maintain water quality but also remove radioactive ions which have leached into the water. At some stage the ion-exchange resins require regeneration and the radionuclides adsorbed therefore reappear in the regenerant liquors. If these are in sufficiently low concentration they may be discharged under authorization, as we shall see in chapter 5, via the condenser cooling water of the power station. If the levels are too high for discharge, provision is made for the use of non-regenerable resins which can be disposed of, in a different manner, as solid waste.

The subsequent fate of the spent nuclear fuel has become the biggest issue in the history of nuclear power development. As stated in the preface, it is not intended to enter into the issues for and against the use of nuclear power, but it is inevitable that at this stage some mention of the major issues involved must be made, the more so because the reprocessing of nuclear fuel accounts, at present, for the most important source of the longer-lived radionuclides introduced into the environment resulting from the use of nuclear power. Arguments presented for a programme of reprocessing irradiated fuel usually centre around its practical necessity for some types of fuel, its desirability with regard to reducing the bulk for long-term disposal, and its economic desirability resulting from the recovery of plutonium and unused ^{235}U from enriched fuel. It has generally been accepted that some types of fuel, particularly magnox, need to be reprocessed because the fuel element cladding deteriorates fairly rapidly if held too long in cooling ponds; although even this argument has been challenged and the suggestion made that storage in gas-cooled chambers, which is a much more corrosion-resistant method, would be a longer-term possibility. There is no doubt, however, that reprocessing reduces the bulk. With regard to the third point, the economic desirability, more than one issue is involved. In a strict monetary sense the justification of reprocessing the irradiated fuel to recover useful uranium has to be balanced against the cost of obtaining the same amount of material from freshly mined uranium at that time. There are deeper issues here, however, which broaden the discussion to a political one. As we have seen, the distribution of uranium ore deposits is very limited globally and thus any user of nuclear power is placed in an economically and politically vulnerable position if it does not have its own uranium resources.

It has been estimated that the recovery and re-use of uranium from a reactor using enriched thermal oxide fuel would meet some 15% of that reactor's fuel reloading requirements. In addition, the plutonium recovered – if used as a replacement for ^{235}U – would fill about another 20% of that reactor's future fuel requirements. Thus one recycle of the irradiated fuel would result in an extra 35% power being generated from the same amount of uranium ore. Again it must be said that the strict economic desirability of this process rests in the relative costs of obtaining the extra 35% from spent fuel as opposed to obtaining it from the raw material. But if the supply of raw material is in anyway restricted, recycling has the obvious attraction of obtaining a much greater proportion of the potential energy from the fuel already in hand. For example the reprocessing of 1000 tonne of irradiated fuel for use in thermal reactors is equivalent, in energy terms, to about 30 million tonne of coal.

If the use of fast breeder reactors is envisaged, the potential is considerably increased; indeed a breeder reactor programme cannot be realized without the reprocessing of thermal reactor fuel to obtain the ^{239}Pu to use on its own, or mixed with ^{235}U, to fuel the reactor. Use of the fast breeder naturally reduces considerably the need to import uranium in the long term, because the ^{235}U-depleted uranium can ultimately be converted into usable fuel. The potential energy savings are very large; some fifty times as much energy can be obtained because virtually all of the uranium content of the original ore of either isotope can be used as a source of energy. This is a very attractive proposition for uranium-importing countries. A nuclear power programme of twenty 1000 MW(e) thermal reactors would give rise to a lifetime's irradiated fuel of about 15 000 tonne. Extraction of the plutonium and uranium from this spent fuel could be used to fuel a further thirty 1000 MW(e) fast breeder reactors throughout their entire life-times.

What, then, are the arguments against reprocessing? First of all its very necessity has been called in to question. Some countries, such as Canada, have no plans for reprocessing fuel. The irradiated fuel elements from the CANDU reactors are at present stored in cooling ponds. If the fuel is never reprocessed it is planned to dispose of it following its encasement in suitable containers. The USA, which has in the past reprocessed nuclear fuel, has also decided to suspend this practice for the time being. Both Canada and the USA have large uranium deposits; the USA also has a considerable investment in ^{235}U enrichment plants and exports enriched uranium fuel elements. There are, therefore, sound and business-like reasons for discouraging the reprocessing of spent fuel by one's customers. Economic issues clearly contain a number of intangible elements, not least being the future demand for nuclear power as a source of energy. A far more tendentious argument against reprocessing has been the raising of political issues over the possibility of the proliferation of nuclear weapons. This possibility arises from the fact that in the reprocessing of nuclear fuel, as we shall see, plutonium is separated and stored separately. The plutonium consists largely of ^{239}Pu and it would be possible to make a crude nuclear bomb with it. Such a weapon would be relatively large, requiring some 8 kg plutonium to form a minimum critical mass, but apparently quite feasible. Plutonium recovered from the blanket of uranium surrounding the core of a fast breeder reactor could contain a higher percentage of ^{239}Pu, which could be used to produce a more efficient nuclear weapon. Indeed the whole reprocessing industry arose initially from the military need to obtain plutonium for weapon manufacture. It was mentioned previously that the future development of fast breeders was complicated by such political issues as the proliferation of nuclear weapons, and it is made the more so when it is envisaged that later generations of fast breeders will use this highly ^{239}Pu-enriched plutonium as fuel. One therefore has the prospect of nuclear weapons grade material being produced in large quantities, being transported nationally and possibly internationally, and thus open to diversion for use by hostile groups. Such a prospect is rather disconcerting and asks much of public attitude and opinion. The proliferation of nuclear weapons can be prevented only by political action, however: all of the detailed plans are already freely available for any determined nation to construct its own reactor, reprocess its fuel and construct its own nuclear weapons; or it could construct them from enriched uranium. It is probable that some countries have already done so. One also has the prospect of inter-nation conflict over other sources of energy, such as oil; and the necessary military escorts which will be required to safeguard the hijacking of the last tonnes of oil, and other diminishing raw materials, in the future.

Regardless of such arguments, whatever their relative merits, nuclear fuel is reprocessed by some countries – notably those without an indigenous uranium supply – and this practice is likely to continue for the rest of the century. What does reprocessing involve? In the first place the irradiated fuel must be transported away from the power station to the reprocessing plant. This is a more complex task than transporting the fuel to the power station for it is now *very* radioactive. As already mentioned, the IAEA has provided a set of regulations for the safe transport of radioactive materials, and those relating to the transport of irradiated fuel fall within what are called type B packages. These must comply with specific tests to ensure that there will be no loss or dispersal of the radioactive contents under very severe accident conditions. When the package contains over 185 TBq (5 kCi) the consignment is termed, modestly, a large source and additional safeguards are required such as the setting up of emergency procedures which take into account the prescribed routes and modes of transport. As an example, the flasks used to carry irradiated fuel elements from Italy, by road, sea and rail to the United Kingdom for reprocessing are constructed as follows. The flask weighs some 47 tonne and contains some 2·6 tonne of fuel elements. The fuel elements are carried in mild steel 'skips' which fit into the flasks and which are also used to hold the fuel in the cooling ponds. The body and lid of the flask are made of mild steel 37 cm thick. Each water-filled flask has to be pressure tested and subjected to a number of integrity tests such as: being dropped 9 m on to a specified target; being dropped 1 m on to a 15 cm diameter mild steel bar, 20 cm long; immersion to a depth of 15 m in water; and being engulfed in a fire at 800° C for ½ h. The dose rate originating from the external surface of the container must not exceed 200 mR h^{-1} for X and gamma-radiation, and its equivalent for beta radiation. And in addition the external surface must not attain a temperature of over 82° C.

The most well known nuclear fuel reprocessing plant in operation today is that situated at Windscale on the Cumbrian coast of the Irish Sea in the United Kingdom. This plant was built and subsequently developed to reprocess natural uranium magnox fuel with a throughput of 1500 tonne uranium per year. For fuel with a maximum rating of 3·5 MWe tU^{-1} irradiated to 3000 MWe days per tonne, this involves an annual throughput of 3 tonne plutonium and 37 EBq (1 GCi) of fission product radionuclides. A flow chart diagram of fuel throughput is shown in figure 4.7. The arriving irradiated fuel elements are held in cooling ponds for at least 130 days to allow further decay of ^{131}I. The elements are then transferred by remote control to a shielded area, or 'cave' where the cladding material is removed by chopping off the ends and splitting it along its length. This immediately results in the release of gaseous fission products, notably ^{85}Kr which has a half-life of 10·7 years. The ^{85}Kr is discharged to the atmosphere. The cladding is cut into strips and stored under water in a shielded silo. The now naked uranium fuel rod is dropped into a vat of nitric

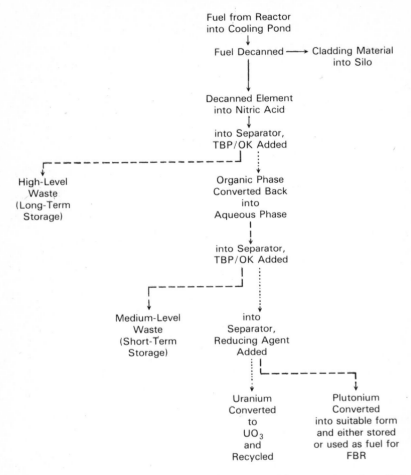

Figure 4.7.   Flow diagram of the reprocessing of irradiated nuclear fuel. Separating processes result in aqueous (‑‑→)and organic (·····→) phases. Low-level wastes arise throughout; principally from condensates and from the washing of solvents. These large-volume, low-level wastes, and water from the cooling ponds, are discharged, under authorization, into the environment.

acid forming a solution of uranyl nitrate at a concentration of about 300 g uranium l^{-1}. This results in the further release of ^{85}Kr. Uranium and plutonium are then extracted from this acid solution by a complex process which involves the use of an organic solvent, tributyl-phosphate (TBP) at a concentration of 20% in odourless kerosene (OK). In the first step an aqueous phase is separated off and this contains high-active waste consisting of the majority of fission product nuclides in the fuel. The organic phase, which contains both the uranium and plutonium from the fuel, is converted back to an aqueous phase and again separated with 20% TBP/OK. The aqueous phase resulting from this separation contains a much lower level of radioactivity and is thus termed 'medium-active' waste. The organic phase, still containing the uranium and plutonium, is now reduced with ferrous sulphamate, which effects the separation of plutonium from uranium. The plutonium is eventually converted to the dioxide form and stored, under high-security conditions, in stainless steel canisters; it can then be used as fuel, and indeed Windscale manufactures the fuel elements for the United Kingdom's prototype fast breeder reactor. The uranium is also converted to powder form and may then be recycled through a fuel enrichment and fabrication plant.

Perhaps the most interesting part of the reprocessing industry is the fate of the high-level waste arising from the first separation stage: its composition will depend on the type of fuel reprocessed and the length of time spent in the reactor. It is an acid solution and, in order to reduce the bulk, it is allowed to boil for a while as a consequence of the heat generated by radioactive decay. The waste from 1 tonne irradiated magnox fuel will be thus gradually reduced to about 40 l high-level waste concentrate. The concentrate is stored in double-skinned stainless steel storage tanks which have seven separate and independent cooling coil systems to remove the heat generated by continued decay of the fission products and actinides. The latter are elements above actinium in the periodic table, radionuclides of which are formed by successive neutron capture. Up to 1976 there were twelve tanks at Windscale containing such high-level waste; eight of these contain 70 m³, and four 150 m³. Such storage is regarded as an interim measure and plans for its long term disposal will be discussed in chapter 7. The medium-level wastes are also stored, but if they decay to an acceptable level they may be eventually discharged, under controlled authorization, to the environment. It will also be clear that throughout the entire operation variable low-level wastes will accumulate – in gaseous, liquid and solid form – and these may also be disposed of in a controlled manner under suitable authorization; principal among these are liquid effluents from the storage ponds, laundry wastes, solvent clean up, evaporator condensates and so on, and a miscellany of contaminated equipment varying from relatively large objects such as glove boxes to small items of glassware.

The above description applies to magnox fuel. Plans are now being made to reprocess uranium oxide fuel at Windscale, and this process will differ from the above in a number of ways, particularly with regard to decladding operations. Unlike the magnox fuel elements, the cladding of oxide fuel elements cannot be removed mechanically, so the entire fuel-pin assembly has to be chopped up. Fuel elements from HTGRs would be even more difficult to reprocess in view of their ceramic nature.

One of the concerns over the reprocessing of nuclear fuel, as already noted, is the production of plutonium which could be misappropriated. Some consideration has therefore been given to the design of fuel-reprocessing techniques which would produce uranium for recycle but leave the plutonium mixed with some uranium and some of the fission product nuclides.

It can be seen from the brief account given in this chapter that the production of power from nuclear reactors results in a variety of radioactive waste. By far the greatest proportion of this waste is that of the spent fuel elements and virtually all of this is retained, either in the stored elements or in the separated high-active waste: it is not released into the environment. Inevitably, however, various gaseous, liquid and solid wastes do arise both at the reactor sites themselves and at reprocessing plants. These wastes are low in radioactive content and it is thus considered acceptable to discharge them into the environment under controlled and monitored conditions. How this is achieved will be the subject of the next chapter. Before leaving the subject of nuclear reactors, however, there is one environmental aspect which should also be mentioned. In common with other forms of power stations large volumes of water are required to provide cooling water for the condensers. Because of the poor efficiency of conversion of thermal energy to electrical energy, about two-thirds of the heat generated is lost to the environment. Environmentally, therefore, the effects of raising the temperature of the receiving water have frequently attracted the greatest attention in many countries. Legislation prohibits the discharge of water at temperatures incompatible with the local environment. For example, in the USA, the approved temperatures for cooling water discharged into tidal waters differ from state to state; in Alaska the water must be less than 20° C whereas in Louisiana the maximum allowed is 36° C. In addition to thermal stress, mechanical damage to aquatic organisms is caused by passing large volumes of water through grids, filters, and the complex plumbing of the condenser cooling system. These problems are not unique to nuclear power reactors, but if the present trend towards larger reactors continues, the capacity of the local environment to absorb waste heat may well become a critical factor. One way of overcoming this is to make use of the otherwise wasted heat. This may be done directly by heating local buildings – an idea which has not proved to be very practical – or indirectly, by using the heated water to rear fish and shellfish at rates of production greater than those which could otherwise be attained. This latter method has been used, with a fair degree of success on a relatively small scale, at nuclear plants in various countries.

5. Radioactive wastes and the public

5.1. *Introduction*

It is impossible to generate electricity from nuclear power without some release of radioactivity into the environment. Not that nuclear power is unique in this respect: the production of electricity from the burning of fossil fuels, particularly coal, results in the release of radionuclides such as ^{226}Ra and ^{228}Th into the atmosphere. The question arises, therefore, as to how much radioactivity one can deliberately introduce into the environment, and at what rate. Of prime concern is the protection of man. The ICRP, which draws up recommendations for those people exposed to ionizing radiations through their occupations, has also drawn up recommendations for the public at large. There is, therefore, an international set of guidelines which can be used to set standards for the safety of the public. It is the responsibility of regulating departments to control the discharges of radioactivity to the environment in such a way that these guidelines are not exceeded. The precise methods used to meet these standards differ somewhat from one country to another, and their differences highlight the complexities of radiological control. All of the methods involve the collection of a large number of data relating to concentrations of radionuclides in the environment, however, and where these data are collected by authorizing departments it should be borne in mind that they are the second set of data obtained because, as will be described in this chapter, the nuclear sites themselves are required to make environmental measurements and these, too, are submitted to the regulating departments.

5.2. *ICRP guidelines for the general public*

The maximum permissible doses discussed in chapter 3 for people occupationally exposed are considered to be upper limits and thus not generally applicable to the public; occupational workers are monitored whereas the public in general obviously cannot be monitored individually. Any recommendations applicable to them must, therefore, also be considered as something of a theoretical concept, and maximum permissible doses would not have quite the same meaning. But one can take a number of steps to ensure that individuals are unlikely to receive more than a specified dose; these steps involve controlling the source from which such an exposure is likely to take place, monitoring likely pathways of exposure to the public, and application of the necessary statistical calculations to such data. Thus instead of maximum permissible doses, the term *dose limit* has been used for the public. The limits recommended by ICRP No. 9 (1966), which gained international acceptance, were an order of magnitude lower than the occupational limits. Thus hands, forearms, feet and ankles were limited to 75 mSv (7·5 rem) per year; skin, bone and thyroid to 30 mSv (3 rem) per year, gonads and red bone-marrow to 5 mSv (0·5 rem) per year; and other single organs to 15 mSv (1·5 rem) per year.

The general public differs from occupationally exposed workers in one important respect; it includes children. An exception to the above recommendations was therefore made by the committee to ensure that the annual dose to the thyroid of children below the age of 16 did not exceed half the adult dose, and was therefore limited to 15 mSv (1·5 rem) per year. For internal exposure, values one tenth of those given in table 3.13, for a 168-h week, were recommended by the ICRP as being acceptable for the general public; again assuming that there was no external source of exposure, a modification factor being applied if necessary.

Although one might consider that those working with radioactive materials – or other sources of ionizing radiations – constitute one set of individuals and all of the rest constitute another, it is not quite as simple as that; particularly where releases of radioactivity into the environment from a point source, such as a nuclear power station, are concerned. It is inevitable that a certain section of the public – those living near the power station, or those otherwise peculiarly related to the potential radioactive releases – are more likely to be exposed than others in the population at large. It is, in fact, usually feasible to define such groups of people, termed the *critical groups*, as distinct from the rest of the public.

With regard to the potential genetic damage resulting from ionizing radiations, however, it is indeed the total population dose which needs to be considered, as it is for radioactive fallout (chapter 2). The commission therefore recommended that a limit of 50 mSv (5 rem) per generation should be used for the whole population for sources other than medical exposures and natural background. This has frequently been interpreted as equivalent to 1·7 mSv (0·17 rem) per year on the assumption that 30 years constitutes a generation time span.

It is to be expected that the more recent recommendations of the ICRP, embodied in their publication No. 26 and discussed in chapter 3, should also apply to recommendations for the general public. A whole-body dose equivalent limit of 5 mSv (0·5 rem) per year for stochastic effects is now recommended for application to individual members of the public; it is equivalent to a mortality risk, from radiation induced cancer, in the range of 10^{-6} to 10^{-5} per year. This limit is expected to result in an *average* dose equivalent of less than 0·5 mSv (50 mrem) to people who are not part of a critical group. The necessary 'weighting factors' relative to this whole-body dose equivalent for the general public are exactly the same as given in table 3.8. To put this figure of 0·5 mSv (50 mrem) into perspective, it will be recalled that in chapter 2 it was shown that the annual internal dose rate for organs resulting from the *natural* radioactivity in the body (table 2.12.) averages to about 0·2 mGy (20 mrad), equivalent to 0·2 mSv (20 mrem) for gamma radiation – the principal source being ^{40}K. It has even been calculated that one would receive an annual dose of about 3 μSv (0·3 mrem) from one's partner's body-burden of ^{40}K by sleeping in a double bed – presumably assuming normal sleeping habits.

An overriding annual dose-equivalent of 50 mSv (5 rem) has now been suggested for any one organ of a member of the public, in order to prevent the induction of non-stochastic effects. In keeping with its general findings that genetic effects are unlikely to be of greater relative importance than somatic effects, ICRP No. 26 does not propose any dose limits for populations as opposed to individuals; this is a considerable departure from previous recommendations.

Nevertheless, it is important to estimate the exposure of the general public in a collective manner, and the doses to which populations will be committed in the future, in order to justify a particular choice of waste disposal. The concepts of collective and committed dose were introduced in chapter 2 in relation to radioactive fallout, but it is necessary to define here the different quantities used, as currently recommended by the ICRP, in the field of radiological protection. Unfortunately, some of them are rather tongue-twisting, and the continual change in word order is very confusing. First of all, in order to assess the dose received by an exposed population, in dose equivalent terms, use is made of the *collective dose equivalent* (*S*). It can be derived either by

$$S = \Sigma_i \overline{H}_i N(\overline{H})_i \qquad (5.1)$$

where $\overline{H}_i$ is the average, or *per caput*, dose equivalent in the whole body (or any specified organ or tissue) and $N(\overline{H})_i$ is the number of individuals of sub group (*i*) in the exposed population. Alternatively, when relating the collective dose to a particular source (*k*) of radiation, the collective dose equivalent can be obtained by integration of the ranges of dose rates within the population, i.e.

$$S_k = \int_0^\infty HN(H)\, dH \qquad (5.2)$$

where $N(H)dH$ is the number of individuals receiving a dose equivalent in the range of *H* to *H* + *dH*.

In order to extrapolate the estimates in time, it is necessary to calculate the dose commitment. The collective dose *rate* will vary as a function of time and thus the *total* collective dose equivalent, from a particular source, can only be obtained by integrating the collective dose equivalent rate ($\dot{S}$). This quantity is termed the *collective dose equivalent commitment* (*S^c*): thus

$$S^c = \int_0^\infty \dot{S}(t)\, dt \qquad (5.3)$$

Both *S* and *S^c* are used in justifying, or optimizing, different choices of waste management practice as will be discussed below. In addition, both definitions can be adjusted to relate to *effective* dose equivalents which, as was discussed in chapter 3, relate to risks of mortality plus a proportion of hereditary effects. One can therefore derive quantities with such unwieldly names as the *collective effective dose equivalent commitment*. One should also note that the collective quantities are frequently expressed as man-Sv (or man-rem), to distinguish them from individual doses.

As if deliberately to complicate the subject for more general students, brief mention must made of yet another quantity, the *dose equivalent commitment (H^c)*. In this case, instead of using the collective dose equivalent rate, an average, or *per caput* dose equivalent rate is derived and integrated in a manner analogous to that of equation (5.3). The particular value of the dose equivalent commitment is that it can be used to estimate the maximum annual *per caput* dose equivalent, from a continuing practice, at some time in the future. In other words it allows comparisons to be made between different situations, or with natural background, at a given point in time. Again it can also be defined in terms of *effective* dose equivalent.

Finally, it will be apparent that for very long-lived nuclides it may be desirable, in certain circumstances, to terminate any of the dose commitment quantities at a specified point of time (T) in the future. The results of such integrations are called *incomplete* or *truncated* dose commitments.

The ICRP has not left the matter there – simply recommending a list of dose limits – but has added a proviso, the wording of which has continually evolved. This proviso in 1955 stated that "...it is strongly recommended that every effort be made to reduce exposure to all types of ionizing radiation to the *lowest possible level*". In ICRP No. 2 (1959) it was recommended that "...all doses be kept *as low as practicable*, and that any unnecessary exposure be avoided". By 1966 the recommendation had been expanded so that ICRP No. 9 recommended "...all doses be kept *as low as is readily achievable, economic and social considerations being taken into account*". In view of different interpretations being made with regard to this last version of the recommendation, ICRP No. 22 (1973) further suggested that "*readily achievable*" be replaced by "*reasonably achievable*". The ICRP No. 26 has, as we shall see, slightly changed the wording yet again.

All of this may seem to be a mere playing with words, but behind it lies one of the biggest factors in the continuing debate over the amounts of radioactivity released to the environment as a result of generating electricity from nuclear power: how much effort and money should be expended to achieve a desired low rate of release? It has been argued that if it is technically possible to bring about further reductions in the amounts released then this should be done; but in order to justify the cost of implementing such technology it is necessary to evaluate what would be saved in terms of 'harm' done to man. As nuclear power is designed to produce electricity one should also take into consideration the benefits gained from its production. In short, therefore, whereas originally the philosophy of radiation protection was dominated by considerations of the relative effects of different types of radiation to organs of differing radiosensitivity and so on, it has now become dominated by what are usually known as cost/risk and cost/benefit factors. In doing so, the release of radioactivity into the environment has come more into line, as it should, with other everyday practices. Such factors are used all the time, particularly in evaluating 'safety' features. For example a cost/benefit analysis study would be made to decide whether the higher expenditure on a manned crossing on a railway line, although safer, is justified relative to the cost of replacing it with an automatic half-barrier system. The equation would be balanced taking into account both their relative expenditures and the number of deaths and injuries expected, using monetary values for the 'cost' of a death or injury. The term 'cost' does not have to be taken literally. In fact, costs are usually considered to comprise the sum total of *all* negative aspects of the equation. Similarly the 'benefits' are considered to include all of the benefits gained by society as a whole. It is a very difficult subject and one which has been dismissed by many critics because of the impossibility of using the same units of measurements in deriving the equation.

Similarly the concept of 'risk' and its acceptance is a much debated subject. Many tables are compiled to illustrate the risks to which we are exposed continually, either at work or in life generally. It is a truism that some occupations, such as mining, are more hazardous than others; and that some sports, such as rock-climbing, are more hazardous than others. Both are choices which can be made by individuals, although it is unlikely that the majority of people have any idea of the absolute level of risks involved. Numerical values are available, but even when given such information it is doubtful if the majority of people would use them. Nevertheless, there is no reason in principle why cost/benefit approaches should not be used with regard to the management of radioactive wastes when they are used in other fields of risk assessment; especially when it is realized that zero risk, and absolute safety, are illusions. In fact, the effects of radiation lend themselves to cost/benefit analyses because they are already evaluated in terms of risk per unit dose, for stochastic and non-stochastic effects, as discussed in chapter 3. One should also not lose sight of the fact that cost/benefit analyses are only applied to considerations of reducing doses to levels even lower than those which are recommended as upper limits for the public in general, and that all of these calculations incorporate numerous safety factors.

In applying the 'as low as reasonably achievable' philosophy to waste management, a method of 'optimization' is recommended by the ICRP. This aims to assess the level of expenditure on protection at which the *total* cost of the waste disposal system is least. The total cost may be considered as a combination of two different sets. One is the direct monetary cost of handling the waste, packaging it, and minimizing its introduction into the environment (set A); the other is the cost of the radiological detriment, which can be converted to a monetary equivalent by assessing the resultant health effects (set B). If these two sets are related to a variable which reflects exposure, for example the collective dose equivalent, or the collective dose equivalent commitment, as in figure 5.1, it can be seen that whereas set A decreases with

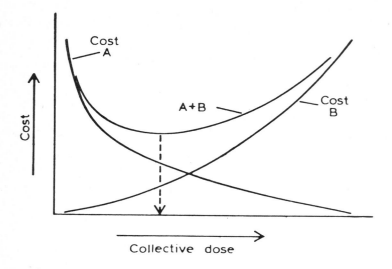

Figure 5.1.   Differential cost–benefit analysis in which 'cost' is related to a variable reflecting exposure—for example collective or committed dose. A is the direct monetary cost, B is the cost of the radiological detriment, and the dotted arrow indicates the minimum value of A + B. (Based on ICRP No. 22).

increased exposure, set B increases. The two sets of 'cost' are additive, and thus the optimization technique attempts to achieve that level of expenditure at which the combined costs are at a minimum. Thus, from an administrative point of view, it is not simply a question of balancing a straightforward cost/benefit equation but of solving a differential equation which attempts to *maximize* the net benefit in relation to the collective dose. It is not always realized that the objective of radioactive waste disposal, as recommended by the ICRP, is not to retain *all* of the waste for *ever* – even if it were technically feasible – because this would not achieve the *optimum* level of protection.

In summary, therefore, the rather cumbersome terminology of ICRP No. 26 states that a system of dose limitation be applied "...to ensure that no source of exposure is unjustified in relation to its benefits or those of any available alternative, that any necessary exposures are kept as low as is reasonably achievable, and that the dose equivalents received do not exceed certain specified limits and that allowance is made for future development''.

Returning again to the whole-body dose equivalent limit of 5 mSv (0·5 rem) per year for individual members of the public recommended in ICRP No. 26, which is expected to result in an average *per caput* yearly dose equivalent of less than 0·5 mSv (50 mrem), it is interesting to compare this number with data presented in chapter 2. The average adult, in a western country, might expect to receive an annual effective dose equivalent of about 1 mSv (100 mrem) from natural background, about 0·01 mSv (1 mrem) from fallout, and about another 0·01 mSv (1 mrem) from various miscellaneous sources. In addition, an average 0·5 mSv (50 mrem) would be expected from diagnostic medical examination using irradiation devices. It is even more interesting, however, to make comparisons with specific examples. The difference in absorbed dose for a person moving from Edinburgh to Aberdeen was mentioned in chapter 2; his annual absorbed dose would virtually double from about 1 mSv (100 mrem) to 2 mSv (200 mrem). Similarly a person in the USA, moving from one of the eastern states to a town in the state of Colorado, could increase his annual absorbed dose from an average 1 to 1·5 mSv (100 to 150 mrem) up to 2·5 mSv (250 mrem) because of a combination of the difference in altitude and the difference in the mineral content of the areas. Even moving house could be significant. A person in Sweden living in a wooden house (24 h a day) would receive an annual absorbed dose rate, in air, of less than 40 μGy (4 mrad), whereas a person living in a house made of brick (24 h a day) would receive an absorbed dose of about 3000 μGy (300 mrad). For gamma and beta radiation these values are equivalent to 40 μSv (4 mrem) and 3000 μSv (300 mrem) respectively; but because much of this absorbed dose results from alpha emitters, the actual dose equivalent values are much higher, alpha particles having an effective quality factor of 20 (chapter 1). One is therefore considering annual differences, in dose equivalents, of the same order as 5 mSv (0·5 rem) recommended as limits by the ICRP. If it really is considered worthwhile to spend money in order to prevent some members of the public from being exposed to an extra, maximum, of 5 mSv (0·5 rem) per year, is it desirable to rehouse people living in brick houses into houses made of wood?

5.3. *Reactor siting*

The first prerequisite for any nuclear power installation is consideration of where the station should be sited. The two prime factors which must be balanced are the needs to build the reactor as close as possible to where the electricity is required, and yet to keep it as far away as practicable from large centres of populations in case of the unlikely event of an accidental escape of radioactivity: the two factors are usually incompatible. In addition, there are a large number of requirements which are essentially similar to those of any other large industrial complex: accessibility to transport and transmission networks, adequate condenser cooling water supplies in the case of power stations, favourable meteorological or other geographic factors, plus consideration of aesthetic and amenity values. Even the substratum must be chosen carefully, not only for fear of earthquakes but because reactors and their containment buildings are so heavy – a typical magnox reactor weighs some 60 000 tonne.

If a site has been considered to be suitable in all other respects, the criteria which have been used to assess its suitability with regard to exposure of the public, in the event of an accidental release of radionuclides, have changed considerably over the years. In the United Kingdom it was originally suggested that circles of 5 and 10 mile radii should be described around a potential site and maximum permissible sizes of populations specified within 10°, and later 30° sectors of these two concentric circles. Such criteria do not allow for the relative safety potential of different reactors, however, and it is probable that future trends will concentrate more on reactor safety design and containment rather than on such inflexible arbitrary approaches. In the United Kingdom a site licence has to be obtained from the Nuclear Installations Inspectorate, which imposes certain radiological restrictions. In addition, authorization has to be obtained from other relevant ministries in order that discharges to the environment can be made. In the USA, the Nuclear Regulatory Commission has required that a reactor be sited within an *exclusion area*, surrounded by a *low population zone* within which the nearest *population centre* – a densely populated area containing more than 25 000 residents – should be at least 30% further than the distance to the outer boundary of the low population zone. The low population zone must be an area in which, if an accidental release of fission product radioactivity did occur, no individual on its outer boundary exposed to the cloud of released radionuclides would receive a whole-body dose in excess of 0·25 Sv (25 rem), or a total radiation dose in excess of 3 Sv (300 rem) to the thyroid from iodine exposure.

There is more than one reason for building nuclear reactors in low-population areas. It is obviously desirable to ensure that one section of the country's population is not placed at a greater risk than others, but a more practical reason is to ensure that the resident population in close proximity to a nuclear power station is small enough to enable appropriate measures to be undertaken on its behalf in case of an emergency. A reactor, as has been said, cannot explode like a nuclear bomb. If a serious accident did occur, however, a cloud of radioactivity could be released. But such a cloud would only travel at the rate of the maximum wind conditions prevailing at the time, and because of the very high temperatures which would probably be associated with such a release there would also be a fair degree of thermal uplift. It is nevertheless desirable that the population at risk should be sufficiently small to be monitored, at the individual level, immediately following such an accident, or even to be temporarily evacuated *en masse*.

With regard to the environment there are, again, differences of detail from one country to another. All operators of nuclear sites not only have to work within the limits imposed upon them by licence but also have to demonstrate that their activities do not result in any 'harm' to the environment around the site. Thus in the United Kingdom, for example, a *district survey* is made, before the station becomes operational, to record the natural radioactivity that is present for future comparisons, The survey needs to be of sufficient magnitude and duration to give an indication of possible seasonal changes. Typically a number of gamma radiation measurements are made at defined points up to a radius of ~8 km, and samples of milk, grass, grass-root mat and soil are collected from farmlands in the area. If sources of drinking water exist in the neighbourhood, samples are measured for total beta activity. Surveys are continued throughout reactor start-up and thereafter throughout the life of the reactor.

Environmental matters in the USA are given even greater consideration. In addition to a Preliminary Safety Analysis Report which has to be submitted to the Nuclear Regulatory Commission, an Environmental Impact Report must also be presented. The latter document must contain a detailed evaluation of the ecology of the proposed site – including any special points of interest, such as species rarity – and a detailed predictive model of an 'impact quotient' supplied which would correlate data obtained on a district survey basis plus a total evaluation of all relevant data which can be provided. Thermal effects also figure largely in such impact reports.

5.4.   *Release of gaseous wastes*

As discussed in chapter 4, gaseous wastes arise in different ways. They consist of noble gases (radionuclides of Kr and Xe), halogens, tritium and tritiated compounds, neutron-activation gases, and particulates. The production of nuclides of Kr and Xe differs markedly from one

TABLE 5.1.   *Effect of delaying fission noble gaseous waste from a light water cooled reactor before discharge to the atmosphere. (The half-lives are those given in the reference, and may not compare with recent estimates.)*

Nuclide	Half-life	Emission rate $GBq\ s^{-1}$ ($Ci\ s^{-1}$)			
		Decay 0 h		Decay 24 h	
^{143}Xe	1 s	18·6	(0·502)		
^{94}Kr	1 s	55·5	(1·50)		
^{93}Kr	2 s	90·3	(2·44)		
^{141}Xe	3 s	100	(2·70)		
^{92}Kr	3 s	150	(4·05)		
^{91}Kr	10 s	112	(3·03)		
^{140}Xe	16 s	90·7	(2·45)		
^{90}Kr	33 s	84·4	(2·28)		
^{139}Xe	41 s	71·0	(1·92)		
^{89}Kr	3·2 m	329	(8·88)		
^{137}Xe	3·8 m	37·7	(1·020)		
^{135m}Xe	15 m	5·85	(0·158)		
^{138}Xe	17 m	16·7	(0·450)		
^{87}Kr	1·3 h	3·77	(0·102)		
^{83m}Kr	1·86 h	0·559	(0·0151)	0·00007	(0·000002)
^{88}Kr	2·8 h	3·57	(0·0966)	0·0094	(0·000254)
^{85m}Kr	4·4 h	1·15	(0·0312)	0·0178	(0·000480)
^{135}Xe	9·2 h	3·32	(0·0896)	0·559	(0·0151)
^{133m}Xe	2·3 d	0·034	(0·00093)	0·0255	(0·000688)
^{133}Xe	5·27 d	0·921	(0·02490)	0·810	(0·0219)
^{131m}Xe	12 d	0·0027	(0·000074)	0·0026	(0·000070)
^{85}Kr	10·4 y	0·0014	(0·000037)	0·0014	(0·000037)

s = seconds; h = hours; d = days; y = years.

From Eichholz, G.G., 1976, *Environmental Aspects of Nuclear Power*, (Michigan, USA: Ann Arbor Science Publishers); courtesy of the author and publishers.

design of reactor to another because they are released principally as a result of failure of the cladding material. In magnox reactors, in which burst fuel cans are quickly recognized and removed, the amounts arising are very small. The majority of fission noble gaseous radionuclides have fairly short half-lives (table 5.1) and thus in PWRs their concentrations in the primary coolant are considerably reduced as a result of radioactive decay. The PWR gaseous waste arises when some of this primary coolant is periodically bled off and treated – degassed – and from general leakage and purging of the reactor containment building. It will be recalled that in BWRs the primary coolant is used to create steam which is used directly to power the turbines. An important feature of this design is an air ejector system which passes steam through a series of nozzles to create a vacuum which removes air from the condenser. In addition, a gland seal system is used to seal the main turbine by passing high pressure steam over a series of ridges. Both of these systems give rise to gases which must be vented from the site, and these contain a high proportion of fission noble gases. Thus, whereas a PWR may release 370 to 740 GBq/MW(e)y (10 to 20 Ci/MW(e)y), a BWR may release well over 37 TBq/MW(e)y (1000 Ci/MW(e)y); i.e. an increase of two orders of magnitude for each megawatt of electrical power produced during a year's operation.

In contrast to the noble gases produced by fission, one important noble gas, ^{41}Ar, arises from the earlier magnox reactors by neutron activation of the stable ^{40}Ar in the shield cooling air. Measurements have indicated a release of up to $11 \cdot 1$ TBq/MW(e)y (300 Ci/MW(e)y). Other neutron activation gases arising from magnox reactors are ^{35}S and ^{16}N: the latter is also formed in the coolant of PWRs. One long-lived neutron activation nuclide, ^{14}C, with a half-life of 5730 years, arises in all types of reactor but more so in those which are graphite moderated. Tritium also arises in all types of reactor, as a result both of ternary fission and neutron activation of Li and B dissolved in, or in contact with, the primary coolant. Of the other gases which are released, only iodine need be mentioned here. A number of short-lived isotopes are produced but it is ^{131}I, with a half-life of 8 days, that is of environmental significance. The much longer-lived ^{129}I, with a half-life of $1 \cdot 6 \times 10^7$ years, has not been found in the vicinity of nuclear power plants, but it is considered to be of some importance with regard to nuclear fuel reprocessing plants.

Because of the delay between removal from the reactor and reprocessing – if that takes place – the shorter-lived nuclides contained in the fuel elements are of little significance in considering the gaseous releases from a reprocessing plant. The largest discharge to the air is that of ^{85}Kr which escapes when the fuel cladding is removed. At Windscale over 37 PBq (1 MCi) is discharged each year.

A variety of methods is used to minimize the discharge of gaseous wastes, and methods are constantly being evaluated with regard to meeting the ICRP's recommendation to reduce the doses to the public to a level that is as low as is reasonably achievable. Gaseous wastes are passed through HEPA filters, immediately followed by a bank of activated charcoal adsorbers. The charcoal is termed 'activated' because it has been heated in an atmosphere of steam to drive off organic matter, thus increasing the number of binding sites; surface areas of up to 1800 $m^2 g^{-1}$ can be achieved. Iodine is usually removed by use of charcoal adsorbers although other methods, such as liquid scrubbing – in which iodine is combined with other chemicals in solution – are also used. The noble gases, being chemically inert, pose special problems. Early designs of BWRs adopted a policy of holding-up the gases for ½ h, to allow a number of the nuclides to decay, before releasing them to the atmosphere via a 100 m stack. More modern designs of both PWRs and BWRs have concentrated on compressing the gases into delay tanks to allow much greater isotopic decay. The effect of only 1-h delay can be seen in table 5.1. Recent advances have introduced a variety of additional means of reducing the gaseous content of reactor effluents. These include gas and vacuum stripping devices, cryogenic distillation, and the catalytic recombination of hydrogen and oxygen. Regardless of the methods used, however, some radionuclides have to be vented to the atmosphere.

The most obvious regulation requirement for a radionuclide released to the atmosphere in a gas might be that it should not be released at a concentration greater than that recommended as an $(MPC)_a$ value in table 3.13, reduced to one tenth for the general public. But this would ignore the enormous dilution which takes place in the atmosphere. If one considers a point source of release, the turbulence of the air flowing past this point source causes the radionuclides to be dispersed; such dispersion, as well as diffusion, being very variable and dependent upon wind speeds, local air temperature gradients, nearby ground contours and large buildings. One therefore has to generalize to a large extent, and a number of mathematical models have been developed, mainly based on Pasquill's equations, which assume that the concentration of a nuclide in a 'cloud' of released gases would be distributed in a Gaussian manner, both vertically and in a cross-wind direction. Measurements are made for an individual site, unless suitable data already exist, on the frequency of occurrence of different types of dispersion, referred to as *stability categories*. The categories are listed A to G according to Pasquill's classification, ranging from extremely unstable to extremely stable atmospheric conditions. By solving the equations for each type of stability category, and using the data on the frequency of occurrence of each type of stability category in that area, a weighted mean, annual, concentration distribution at ground level can be derived for each radionuclide.

The radionuclides released will give rise to both external and internal irradiation of man. Radionuclides of the inert noble gases, particularly [41]Ar and [85]Kr, will cause irradiation of the skin by both beta and gamma radiation, and external radiation of the gonads by gamma radiation. The results of theoretical calculations on the dose to gonads

TABLE 5.2. *Gonadal dose from noble gases released in airborne effluents from reactors. The doses are normalized to a release rate of 37 GBq (1 Ci)*

Type of reactor[†]	Dose per unit release 10^{-11}Gy (10^{-9} rad)			Collective dose per unit release[‡] 10^{-5}man-Gy (10^{-3}man-rad)	
	1 km	10 km	100 km	1-10 km	10-100 km
BWR	44	2·0	0·015	0·16	0·29
PWR	5·7	0·3	0·008	0·02	0·07
GCR	86	3·2	0·006	0·26	0·22

[†]Effective stack height of 30 m, and average weather conditions.
[‡]Assuming a population density of 100 km^{-2}.

Data from the Report to the United Nations Scientific Committee on the Effects of Atomic Radiation, 1977, General Assembly document 32 Session, Supplement No. 40 (A/32/40) (New York: United Nations).

TABLE 5.3. *Relative contribution of the major noble gases, in percentages, to the gonad values given in table 5.2.*

Type of reactor and composition of gases	Fraction of total release (%)	Distance from reactor (km)		
		1	10	100
BWR				
^{135}Xe	26	15	17	55
^{88}Kr	14	32	39	23
^{88}Rb[†]		2·6	12	7·4
^{133}Xe	14	0·7	1	6·1
^{138}Xe	13	21	4·7	
^{138}Cs[†]		3·5	8·7	0·1
^{87}Kr	12	18	14	1·3
PWR				
^{133}Xe	90	37	48	74
^{135}Xe	3	13	15	12
^{138}Xe	1	13	2·7	
^{88}Kr	0·9	16	19	2·8
GCR				
^{41}Ar	100	100	100	100

[†]Daughter products, formed after release and decay of parent nuclide.

Data from the Report to the United Nations Scientific Committee on the Effects of Atomic Radiation, 1977, General Assembly document 32 Session, Supplement No. 40 (A/32/40) (New York: United Nations).

at different distances from a release point are given in table 5.2. The table shows both the estimated dose and the collective dose per unit release. The individual dose may decrease with increasing distance from the reactor but the collective dose may increase because there is a greater number of people exposed. The individual doses received are extremely small indeed. The relative contribution of the individual nuclides to these absorbed doses for different types of reactor are given in table 5.3. A number of measurements which have been made in the vicinities of nuclear power stations largely substantiate these theoretical calculations.

Continual surveillance is made of airborne radioactivity. Three types of sample may be required: total air, particulates and precipitation. In order to measure all possible radionuclides present in the air would require a complex array of instrumentation; such measurements are usually unnecessary. A feature of the environs of nuclear power stations in the United Kingdom are hanging lampshade-looking devices called *tacky shades*. These are suspended at ~3 m above the ground and act as passive collectors of particulate airborne activity. The shades are made of a muslin cloth impregnated with a sticky non-setting resin. They are routinely collected and analysed. In the USA horizontal, gummed acetate devices are used to collect particulate matter. Other devices such as rainwater collectors and dry deposition collectors are also laid out on a grid basis around each site, and the routine analysis of their contents is part of the task of the district survey team and of the authorizing departments.

For a number of radionuclides the concentrations which are attained in air, or the absorbed dose in air from their deposition on the ground, are of secondary importance. The prime concern is that they are passed along a variety of food chains and thus result in a source of internal radiation exposure to man. The mathematical models which have been used to describe gaseous dispersion apply to radionuclides which are particulates, or droplets, of less than 20 μm diameter. If one assumes that such particulates have a density similar to that of water, they will not be immediately deposited under the direct influence of gravity. Estimates can therefore be made of where such deposition will occur as the result of the 'cloud' of released gases reaching ground level.

Of the nuclides present in a typical composition of gaseous releases from a power plant, ^{131}I has received the greatest attention because it appears in cow's milk. Terrestrial food chains important in human nutrition are generally much shorter than aquatic ones; and that for iodine is very short indeed. Pastures contaminated by ^{131}I are directly grazed by cows and the ingested iodine transferred to milk. Complications do arise, however, because the ^{131}I discharged may be present in a number of forms – particulate, elemental, organic, or as hypoiodous acid. In elemental form iodine is readily deposited, as is the particulate form. Organically bound iodine, however, is deposited at only one thousandth of the rate of elemental iodine. The behaviour of hypoiodous acid is uncertain. From a regulatory point of view the most useful calculations are those which relate the concentration of ^{131}I per unit area of pasture land to a concentration of ^{131}I per unit volume of milk produced. Measurements have been made of the area grazed by single cows per unit time – a parameter which appears to be relatively more constant than estimates of food intake per unit time – and average transfer coefficients derived which present the ^{131}I transferred to a litre of milk as a percentage of that taken up by the cow per day. In controlled experiments the average transfer coefficient has been calculated to be between 0·5 and 1·0%. Relating these values to contaminated pasture, continual grazing over an area containing 37 kBq (1 μCi) of [131]I m^{-2} would result in a milk concentration of approximately 7·4 kBq (0·2 μCi) of [131]I l^{-1} milk. In practice such calculations usually produce results which are higher than those observed directly, partly because herds may be fed supplementary foodstuffs even during the summer months.

The direct contamination of fresh fruit and vegetables has also to be considered. For short-lived nuclides such as ^{131}I, accumulation via the root system of plants is of little importance; but it does have to be considered for longer-lived radionuclides, as does the possible transfer from the primary foodstuff, such as milk, to secondary products such as butter and cheese. This is of greater relevance for gaseous releases from fuel reprocessing plants which may contain ^{137}Cs and ^{90}Sr; both are rapidly incorporated into cow's milk and consideration has therefore to be given to their continual incorporation into herbage. Transfer coefficients for the direct transfer of ^{137}Cs to cow's milk are of the same order as that for ^{131}I; but for ^{90}Sr values ranging from 0·05 to 0·22% have been reported.

All of these calculations are used in setting limits on gaseous discharges from both nuclear power plants and from nuclear fuel reprocessing plants. To give an example of how these limitations work in practice, table 5.4 gives the estimated dose equivalents resulting from the atmospheric discharges from the Windscale reprocessing plant – which emits a wide range of nuclides – to the surrounding population. The data relate to 1976 and are for high stack releases only. It can be seen that none of the estimated doses received exceeded 1% of the recommended limits for public exposure.

TABLE 5.4. Estimates of committed dose equivalent resulting from atmospheric discharge from high stacks (~100 m) at the Windscale reprocessing plant in 1976.

Nuclide	Organ	Annual dose limit μSv (mrem)	Route	Committed dose equivalent μSv (mrem)		
				At 200 m	At 1 km	At 5 km
^{90}Sr	Bone marrow	5000 (500)	Milk	4·6 (0·46)	1·5 (0·15)	0·38 (0·038)
	Endosteal cells	15000 (1500)	Milk	14·0 (1·4)	4·6 (0·46)	1·2 (0·12)
^{137}Cs	Whole body	5000 (500)	Milk	2·2 (0·22)	0·72 (0·072)	0·18 (0·018)
	Whole body	5000 (500)	Deposited activity	—	1·3 (0·13)	0·32 (0·032)
^{131}I	Thyroid	15000 (1500)	Milk	21 (2·1)	6·6 (0·66)	1·7 (0·17)
^{85}Kr	Gonads	5000 (500)	External γ	—	0·43 (0·043)	0·085 (0·0085)
	Skin	30000 (3000)	External β and γ	—	25 (2·5)	9·4 (0·94)
^{3}H	Whole body	5000 (500)	Inhalation and ingestion	—	0·4 (0·04)	0·2 (0·02)
^{14}C	Whole body	5000 (500)	Inhalation and ingestion	—	0·28 (0·028)	0·25 (0·025)
	Body fat	15000 (1500)	Inhalation and ingestion	—	1·3 (0·13)	1·1 (0·11)
^{129}I	Thyroid	15000 (1500)	Inhalation and ingestion	—	1·2 (0·12)	1·1 (0·11)
Total α activity (^{239}Pu)	Lung	15000 (1500)	Inhalation	—	4·5 (0·45)	1·6 (0·16)
	Liver	15000 (1500)	Inhalation	—	2·9 (0·29)	0·99 (0·099)
	Endosteal cells	15000 (1500)	Inhalation	—	14·0 (1·40)	4·7 (0·47)

From Bryant, P.M., 1978. In *Seminar on Radioactive Effluents from Nuclear Fuel Reprocessing Plants*, Karlsruhe, C.E.C. p.247. Courtesy of the author and of the National Radiological Protection Board.

Gaseous releases are not, therefore, considered to be a significant hazard of nuclear power station or reprocessing plant operation. But there are some areas of uncertainty. Very long-lived radionuclides are able to achieve widespread distribution in the environment. Of particular interest are ^{14}C and ^{129}I, although both ^{3}H and ^{85}Kr are also of interest because these, too, become globally dispersed. The difficulty is not so much a scientific, as a philosophical one. The long-lived radionuclides such as ^{14}C and ^{129}I are being continually produced and will exist in the biosphere for a very long time. The world's population will also increase. Where does one draw the line? This problem is particularly acute when attempting to balance an equation on how much should be spent to reduce the present levels of discharge – if at all – against any conception of benefit. It all depends on the size of the collective dose commitment, or truncated collective dose commitment, the latter depending upon where one draws the line in the future.

There are some fairly obvious limits which one could apply to such calculations. The limit to the world's population is considered to be in the region of 10^{10}. One can similarly make a guess at the length of time nuclear power derived from fission is likely to obtain, say 500 years. A third factor is whether the released radionuclides will remain in the biosphere. It is assumed that both ^{129}I and ^{14}C will mix with, and be diluted by, their stable element reservoirs. Using data which are available it has been calculated by UNSCEAR (1977) that the incomplete collective dose commitment over this time period is about 0·005 man-Gy (0·5 man-rad)/MW(e)y to the thyroid from ^{129}I and 0·03 man-Gy (3 man-rad)/MW(e)y to the bone marrow from ^{14}C. These values can therefore continually be applied to current estimates of the future production of energy from nuclear power.

The long-term fate of one particular element, plutonium, has been a subject of considerable interest. The nuclides of plutonium are particularly hazardous if inhaled, but less if ingested. Discharges direct to the atmosphere are therefore reduced to an absolute minimum but even discharges of plutonium in aquatic wastes may result in an inhalation pathway. It is possible that plutonium discharged into the sea could become airborne, or that contaminated marine sediments could be wind-blown. Both are possibilities currently being investigated. The amounts likely to accrue in the atmosphere from either of these two routes are expected to be extremely small, however, as indicated by measurements made in some coastal areas, and thus it is the long-term fate of environmentally dischargd plutonium which is of interest.

5.5. *Release of liquid wastes*

Low-level liquid effluents arise from the day-to-day operations of nuclear reactors and reprocessing plants; their release to the environment can result in the exposure of man to radiation from a variety of sources. If the release is to freshwater then the most obvious means of exposure is that of drinking water. Where the effluents are discharged into estuaries or the sea, however, this pathway obviously does not obtain, but there are a number of possible means of exposure. In estuaries, where large areas of sand and mud frequently become exposed at low tide, the adsorption of radionuclides onto these sediments presents a source of external irradiation. The consumption of contaminated foodstuffs also provides a means of exposure in all aquatic environments, but particularly for coastal areas. For small occupational groups such as fishermen, the handling of contaminated fishing tackle can also provide a source of exposure. Limitations are therefore made by the relevant authorizing departments, using the ICRP recommendations, on the composition and total amounts of radionuclides which can be discharged as low-level liquid waste. A variety of control procedures has been used; they differ markedly one from another – although all attempt to safeguard the public. The different approaches are worth examining in detail.

The most simple approach would be that of ensuring that individual nuclides in the liquid effluent are not in excess of those recommended by the ICRP as $(MPC)_w$ values for the public. This method, or modifications of it, has indeed been used by a number of countries and is referred to as the *point of discharge control* approach; it has been used in the USSR and in the USA. Its attraction was that, by placing the emphasis on concentration limits at the point of discharge into the environment, it was easily applied by the operators of the plant. The exact details of its application differed from country to country.

A rather different method has been the *specific activity* approach which, instead of taking the ICRP recommendations on the intake of radionuclides as a starting point, took the ICRP data relating to the maximum permissible body burdens of each radionuclide. For the general public, as with the MPC values, the ICRP recommended that 10% of the value for occupational exposure be used. Data on the concentrations of stable elements in the individual organs of the human body were then used to calculate maximum permissible specific activities for each organ, i.e. the concentration of the radionuclide per unit of the stable element as $Bq \ g^{-1}$ (or $\mu Ci \ g^{-1}$); for example x Bq $^{137}Cs \ g^{-1}$ Cs.

The basic premise was then made that, if discharges to an aquatic environment were regulated such that the specific activities of the radionuclides in the receiving water were held below those permitted for man, then they could not be exceeded at any point in the food chain, or in man. The rate of introduction was then set such that this ratio of each radionuclide to stable element in the receiving water was not exceeded. The *specific activity* approach (figure 5.2), as it is termed, is at first sight very attractive. It apparently eliminates the need for speculation on the extent to which any aquatic organism might concentrate radionuclides from the water; and it also eliminates the need for apportioning some fraction of the total intake of radionuclides to marine foods. The method has largely been used in the USA, but despite the apparent attractions it has been beset by difficulties. In the first instance it is not as widely applicable as it appears: it relates only to metabolized elements for example, and does not take into account those nuclides which are taken in orally but are not absorbed – for many radionuclides the gastrointestinal tract is the critical organ for this very reason. It also fails to take into account external dose-rates and thus additional measures have to be taken to safeguard this route

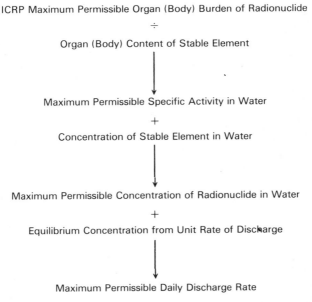

ICRP Maximum Permissible Organ (Body) Burden of Radionuclide

÷

Organ (Body) Content of Stable Element

Maximum Permissible Specific Activity in Water

+

Concentration of Stable Element in Water

Maximum Permissible Concentration of Radionuclide in Water

+

Equilibrium Concentration from Unit Rate of Discharge

Maximum Permissible Daily Discharge Rate

Figure 5.2. The specific activity approach to the discharge of aqueous radioactive wastes.

of exposure. One of the more limiting factors is that it can only be applied with some degree of confidence to marine situations because the chemical composition of sea water is relatively stable; for freshwaters there can be very large differences in the concentrations of some of the major elements such as calcium and potassium – chemicals similar to strontium and caesium, respectively – from one location to another. But even for sea water the precise chemical composition is still a matter of some debate for many of the trace elements, and new methods of analyses are continually revising previous ideas. Many elements are still extremely difficult to measure precisely and there is the added difficulty that for the transuranium elements – neptunium, plutonium, americium and curium – there are no naturally occurring counterparts. The specific activity approach also implicitly assumes that the radionuclide will be in precisely the same chemical form as the stable element already present, and thus behave in a similar manner. From the definition given of a radioisotope in chapter 1 this would appear to be irrefutable, since it was stated that isotopes of the same element differ only in their nuclear composition, not in their number of orbital electrons. But not all chemical forms of an element readily exchange with one another. This is especially true of those elements which may become complexed to organic molecules. Thus if a radionuclide is discharged in a chemical form different from that of the same chemical in the receiving water – and it is likely that there will be a number of different chemical forms already present – the radionuclide may be accumulated by living or non-living material to a greater or lesser extent than the stable element. In fairness, this possibility was, in fact, recognized by the proponents of the specific activity approach and it was suggested that an arbitrary factor of 10 reduction in the calculations be applied to safeguard against it happening. All of these deficiencies may, in part, be balanced by the conservative assumption of the specific activity approach – that all of an individual's food supply is derived from the receiving waters, which is unlikely to be the case. Nevertheless, it is evident that because of the large number of unknowns, and the possibility of errors, it cannot be considered to have been an entirely satisfactory method.

A third approach to setting limits for aquatic discharges, suggested in ICRP No. 7 (1966), recognized the fact that although there will be a number of pathways by which man may be exposed to discharged radioactivity, it will be found that in practice at any given site one, or perhaps two, will prove to be so limiting that if exposure to the public along these pathways is to be kept within ICRP-recommended dose limits, all other exposure pathways will be relatively minor. Thus, for example, it will be found that although a variety of fish species may be caught in any one area, one or two predominate above the others. It is also likely that these species may form a large fraction of the diet to a small section of the community. Similarly, although a large spectrum of radionuclides may be discharged, only a few will predominate in the edible portions of marine animals. Thus one can delineate a few critical radionuclides which are transferred along a few critical pathways, and of these usually one combination predominates above all others. This method of discharge limitation was therefore called the *critical pathway approach*; it has been applied in many countries, but notably in the United Kingdom, where it was largely pioneered.

Unlike the previous methods, the critical pathway approach does require a considerable amount of investigative effort, and a number of quite different pieces of information. The general steps involved are outlined in figure 5.3. Before a nuclear site begins operation a provisional assessment is made of the probable effluent composition and the potential pathways of exposure. Predictions are made of the concentrations which are likely to obtain in the receiving water per unit

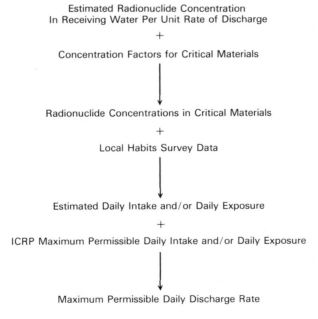

Estimated Radionuclide Concentration
In Receiving Water Per Unit Rate of Discharge

+

Concentration Factors for Critical Materials

Radionuclide Concentrations in Critical Materials

+

Local Habits Survey Data

Estimated Daily Intake and/or Daily Exposure

+

ICRP Maximum Permissible Daily Intake and/or Daily Exposure

Maximum Permissible Daily Discharge Rate

Figure 5.3. The critical pathway approach to the discharge of aqueous radioactive wastes.

rate of discharge, for example at 37 GBq (1 Ci) d⁻¹. This requires a good estimate of the turnover of water at the site of discharge based on hydrographic data. A further estimate must be made of the degree of accumulation of the radionuclides over the receiving water by aquatic organisms which are likely to provide a pathway back to man. This degree of accumulation is expressed as a *concentration factor* defined as Bq (or pCi) g⁻¹ wet of the organism divided by Bq (or pCi) g⁻¹ of the receiving water for each radionuclide. The concentration factor does not necessarily imply that the organism concentrated the radionuclide direct from the water: it may have accumulated it via a tortuous aquatic food chain. Nevertheless, by whatever means the radionuclide was accumulated, its concentration in the organism will probably attain a fairly constant value relative to that of the water, and this can be conveniently expressed as a concentration factor. The concentration factor data are obtained in some instances by stable element analyses but, of course, as with the specific activity approach, there are difficulties in attaining accurate detection levels for many elements. Fallout studies have also yielded many useful data. A reasonable degree of latitude can usually be allowed in these initial estimates, but one very important aspect is the establishment of the working, eating and recreational habits of the local population – and in certain instances of populations at some distance from the site – in order to make quantitative estimates of aquatic food intake, and hours spent on possible contaminated mud and sand banks. This exercise consitutes a *habits survey*. Given all of these data one can then piece together the critical pathways, and make estimates of the daily rate of intake of each radionuclide – or external exposure to it – resulting from a unit rate of discharge. The rate of discharge which, in theory, would result in the defined critical groups of the public being exposed at the ICRP-recommended dose limit has been referred to as the *limiting environmental capacity* for that particular site. It would clearly be unwise to use this value as an upper limit to the rate of discharge, because until the power station actually starts operating its validity cannot be ascertained. An early approach was to apply an arbitrary safety factor, typically a factor of 10 – the so-called *stipulated environmental capacity* – within which the authorized limit was set according to need. After the station had been operating for a while the environmental capacity was then re-assessed on a more reliable basis. Experience would also have been gained on the station's discharge requirements and a revised rate of discharge subsequently specified by the regulatory departments concerned.

Having set the maximum permissible discharge rates, it is also necessary to devise a means of checking one's theoretical values. It would clearly be tedious to go back to first principles every time, and the most practical method has been found to be that of using the value in the calculations which relates to the concentration of the radionuclide(s) in the critical material(s). Thus the calculations may have shown that a particular species of fish, shellfish or area of sediment should not exceed a certain concentration for a specific radionuclide, and the samples can then be routinely collected and analysed to show that this is so. The maximum permissible concentrations in these materials that could be permitted – without exceeding ICRP-recommended dose limits – constitute the *derived working limits* (DWLs).

As a number of concepts have been introduced in describing the critical path approach it would be as well to give a working example of how it has operated in the past. One of the more interesting sites is that of the nuclear power station at Bradwell in England. This twin-magnox reactor station is situated on an estuary, part of which is used for the cultivation of oysters. Local habits surveys confirmed that the consumption of oysters was the principal sea food chain from the area and that some local people consumed up to 30 g oyster flesh daily. An assessment of the radioisotopic composition of the liquid waste to be discharged revealed that ^{65}Zn was likely to be important; the more so when it is known that oysters have a special affinity for zinc. In fact they concentrate zinc to a level some 10^5 times that of zinc in the ambient water. Further research showed that a discharge rate of 37 GBq (1Ci) d^{-1} of ^{65}Zn into the tidal reach was likely to result in an equilibrium sea water concentration of 5·55 mBq (0·15 pCi) of ^{65}Zn g^{-1} sea water. With a concentration factor of 10^5 the oysters could therefore attain concentrations of 0·56 kBq (15 nCi) ^{65}Zn g^{-1} wet; and a consumption of 30 g oysters per day would result in a ^{65}Zn intake of 16·7 kBq (0·45 μCi) d^{-1}. The ICRP maximum permissible intake of ^{65}Zn for the general public – the whole body being the critical organ – was approximately half this figure, at 8·14 kBq (0·22 μCi) d^{-1}. (This value was derived, as explained in chapter 3, by calculating a daily maximum permissible intake on the basis of the (MPC)$_w$ for the public and a drinking rate of 2·2 l d^{-1}, i.e. 10^{-4} μCi ml^{-1} × 2200 ml = 0·22 μCi.) The limiting environmental capacity would therefore be met at a discharge rate of 18·5 GBq (0·5 Ci) d^{-1}, 6·7 TBq (182 Ci) y^{-1}; and the first authorization given limited the discharges from Bradwell to 1·85 TBq (50 Ci) for total radioactivity with not more than 0·185 TBq (5 Ci) of ^{65}Zn, the latter being 1/36th of the estimated environmental capacity. Special ion-exchange plant was

TABLE 5.5.   *Concentrations (wet weight) of radionuclides in the flesh of oysters (Ostrea edulis) from the Blackwater Estuary, UK.*

Radionuclide	Power station outfall		1/3 mile from outfall	
	$mBq\ g^{-1}$	$(pCi\ g^{-1})$	$mBq\ g^{-1}$	$(pCi\ g^{-1})$
^{55}Fe	29·2	(0·79)	2·5	(0·067)
^{60}Co	13·0	(0·35)	1·1	(0·031)
^{65}Zn	1595·0	(43·10)	164·3	(4·44)
^{110m}Ag	14·1	(0·38)	2·6	(0·070)
^{137}Cs	6·3	(0·17)	1·5	(0·041)
^{32}P	12·2	(0·33)	4·4	(0·12)

From Preston, A., and Jefferies, D. F., (1969). In *Environmental Contamination by Radioactive Materials*, IAEA Symposium 117, p.183; Crown copyright, reproduced by permission.

installed at the power station to remove the bulk of ^{65}Zn before discharge via the cooling water outlet. The actual concentrations of ^{65}Zn, and other radionuclides, measured in oyster flesh several years after the station commenced discharge is shown in table 5.5.

The importance of ^{65}Zn was thus confirmed and this radionuclide initially dominated the situation. Over the first few years of operation analyses of the effluent indicated that most of the radioactivity present was due to ^{35}S and ^{3}H, but their impact on exposure to man was negligible due to their lower radiotoxicity and different environmental behaviour. As the operation of the power station reached equilibrium, however, the composition of the effluent gradually changed and ^{134}Cs, and the fission product ^{137}Cs, increased in significance. Caesium is not accumulated by oysters to any great extent (~ 10 times the water value) but another nuclide, ^{110m}Ag which, like zinc, is highly accumulated, gradually increased in importance. Repeated habit surveys also revealed that some individuals consumed as much as 75 g oysters per day. Revised environmental capacities were therefore drawn up (table 5.6). The derived working levels were set at 32·6 Bq (880 pCi) g^{-1} wet and 107 Bq (2900 pCi) g^{-1} wet for ^{65}Zn and ^{110m}Ag, respectively. The concentrations observed in local oysters are far less than even these revised values (table 5.6) and therefore the estimated exposure to the public, even at a consumption rate of 75 g oyster flesh a day, every day, results in a value of less than 1% of that recommended by the ICRP for members of the public.

It is worth considering for a moment what these values mean. The stipulated environmental capacities incorporate safety factors in their

TABLE 5.6. *Environmental capacities and actual radiological importance of the oyster-to-man critical pathway in the Blackwater Estuary, UK.*

Radionuclide	Environmental capacity† in Bq (Ci) based on:		Mean observed concentration (wet weight) in oyster flesh mBq g^{-1}(pCi g^{-1})	Radiological importance (as % of ICRP recommended dose limit)	
	GI tract exposure	Total body exposure		GI tract exposure	Total body exposure
^{65}Zn	1·37 × 10^{13} (3·7 × 10^{2})	7·0 × 10^{12} (1·9 × 10^{2})	18·5 (0·5)	0·009	0·017
^{110m}Ag	1·2 × 10^{12} (3·2 × 10^{1})	2·7 × 10^{14} (7·4 × 10^{3})	55·5 (1·5)	0·17	0·001
^{137}Cs	2·1 × 10^{17} (5·6 × 10^{6})	5·2 × 10^{15} (1·4 ×10^{5})	3·7 (0·1)	<0·001 — 0·18	<0·017 — 0·035

†Assuming consumption rate of 75 g oysters day^{-1}.

From Mitchell, N.T., and Jefferies, D.F., 1973. In *Environmental Behaviour of Radionuclides Released in the Nuclear Industry*, IAEA Symposium 172, p.633; Crown copyright, reproduced by permission.

calculations. These, in turn, are based upon the ICRP-recommended levels for the public which are 1/10th of the levels considered acceptable for persons working in a suitably monitored environment for 168 h a week, 50 weeks a year for 50 years. These figures were derived from data based on high doses of exposure and thus, again to be on the safe side, incorporate one or more safety factors of 10 in them. The estimated exposures of the oyster-eaters, as a result of the Bradwell discharges, are therefore orders of magnitude below any exposure known to cause 'harm' for the basis of the ICRP calculations.

Where nuclear power stations are situated on the shores of enclosed bodies of water the capacity of the water to receive discharged liquid wastes safely has to be very carefully considered. Such an example is that of Lake Trawsfynydd in Wales where another twin-magnox power station is situated. Pre-operational assessments suggested consumption of trout (*Salmo trutta*) as a likely candidate for the limiting critical path, and neutron activation products – ^{32}P, ^{65}Zn, ^{60}Co and ^{124}Sb – were predicted as the nuclides in the effluent which would probably be the most important. Initial discharges were minimal but it soon transpired that the predominant nuclides in the effluent – as has occurred at most of the magnox power stations – were those of caesium, ^{134}Cs and ^{137}Cs. Freshwater fish have high concentration factors for caesium, of the order of 10^3. This value is attained by accumulation from contaminated prey living on the lake bed. A considerable research effort was expended on attempting to relate a number of materials – the water, lake sediment, fish – to the effluent over a longer time-span. It was found, for example, that although the half-time for the water in the lake was estimated to be approximately 50 days – freshwater from a number of streams displacing the water of the lake via the lake's dam – it took approximately 2 years for the ^{137}Cs in the fish flesh to attain an equilibrium value with the input of ^{137}Cs into the lake. The derived working levels were initially set at 16·3 Bq (440 pCi) g^{-1} wet for ^{137}Cs and 7·4 Bq (200 pCi) g^{-1} wet for ^{134}Cs. Trout flesh actually contains only a very small fraction of these values.

The Bradwell and Trawsfynydd examples are now largely of historical interest because little use is currently made of the notion of a stipulated environmental capacity; this is largely because in a very few cases have the discharge needs of a power station been equivalent to public exposure at more than a very small proportion of the ICRP-recommended dose limits. Thus, provided that these needs are very low, the margin between them and the limiting environmental capacity is more than adequate to cover uncertainties inherent in the latter. In fact the trend is now to focus more and more attention on the justifiable needs that a power station has in terms of waste disposal, and in so doing develop more objective means by which the limits set in authorizations can be judged to have been properly optimized. There are some power station sites where actual critical pathways cannot immediately be recognized. Potential pathways are therefore monitored, such as any food species taken in the area, and dose-rate measurements made of any exposed sediment at low tide. At many sites use is also made of biological indicator materials. Algae have been found to be particularly useful in this respect because they adsorb a number of radionuclides from the water. Surveillance of all such materials is routinely made until a critical path emerges, if it does so at all.

The liquid effluent discharge requirements of a reprocessing plant are of necessity greater than those of power stations, but they are controlled by precisely the same methodologies. Indeed the critical path approach which has been applied to Windscale has become a classical example of this method. A large range of radionuclides is discharged to the Irish Sea at Windscale, under authorization, via a pipeline some 2·5 km long. The Irish Sea supports an important commercial fishery, but of all the potential pathways back to man it was found, as a result of carefully controlled experimental discharges, that the most critical pathway was likely to be one which involved an edible seaweed, *Porphyra*. This seaweed was harvested regularly along the coast adjacent to the reprocessing plant. It was not consumed locally, however, but transported some 500 km to an area of South Wales where, together with *Porphyra* collected in other areas of the United Kingdom, it was processed into a foodstuff called laverbread. *Porphyra* accumulates a number of radionuclides from sea water in the vicinity of Windscale – ^{144}Ce, ^{95}Zr/^{95}Nb, ^{90}Sr, ^{239}Pu, ^{241}Am – but particularly ^{106}Ru (figure 5.4). The critical population turned out to be rather large; at least 26 000 people were estimated to eat laverbread regularly. The maximum rate of laverbread consumption determined from initial local surveys was 75 g day^{-1}. Subsequent surveys identified a sub-group of enthusiasts for this particular delicacy, with individual consumption rates as high as 388 g day^{-1}, and a median value of 160 g day^{-1}. This group, of 170 adults of both sexes, was thus found to be more important than the average consumer.

The critical organ for ingested ^{106}Ru, and ^{144}Ce and ^{95}Zr/^{95}Nb, is the lower large intestine; for ^{90}Sr, ^{239}Pu and ^{241}Am it is bone. Because more than one radionuclide is present in *Porphyra*, calculations of DWLs were made for the individual radionuclides. The overall

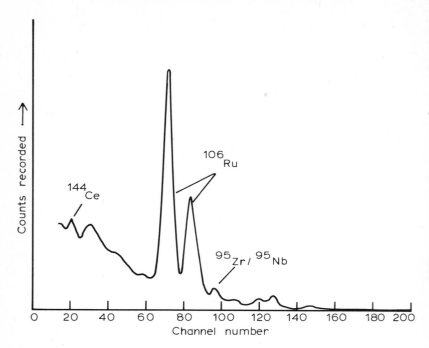

Figure 5.4. Gamma-ray spectrum of *Porphyra* collected near Windscale: analysis was made using a NaI(Tl) crystal and a 200-channel analyser.

discharge limit was then derived by expressing each radio-nuclide's discharge as a fraction of its authorized rate, and ensuring that when all of the fractions were added up they did not exceed unity. The ICRP-recommended annual dose limit for the lower large intestine was 15 mSv (1·5 rem). Estimated doses received by the sub-group for the years 1962 to 1967 ranged from 4 to 7 mSv y⁻¹ (0·4 to 0·7 rem y⁻¹). In fact, these estimates were considered to be conservative because they were based on the assumption that all of the *Porphyra* used in laverbread production originated from the coastline near Windscale, which it did not, and also they did not allow for the incorporation of water – which serves to dilute the radionuclide's concentration – during processsing. The conservative nature of these assumptions was borne out by frequent analyses of commercially obtained laverbread.

    With such a large laverbread-consuming population it was also necessary to take account of the genetically significant dose. The mean acceptable annual GSD was calculated on the assumption that the

TABLE 5.7. *Examples of the critical path approach to the aquatic discharge of low-level waste.*

Site	Critical Group	Material	Radio-nuclide	Organ	Percentage of ICRP dose limit received by critical group	Year
Clinch River (Tennessee, USA)	Clinch River residents	Drinking water	^{90}Sr	Bone	30†	1968
Columbia River (Washington State, USA)	Local fishermen	Flesh of local river fish	^{32}P	Bone	12	1966
Windscale (Cumberland, UK)	Laverbread eaters in South Wales	*Porphyra* seaweed	^{106}Ru	Gastro-intestinal tract	40	1962-67
Bradwell (Essex, UK)	Oyster fishermen	Oyster flesh	^{65}Zn	Total body	0·26	1967-68
Trawsfynydd (Merioneth, UK)	Trout fishermen	Trout flesh	^{137}Cs ^{134}Cs	Total body	3	1968
Chalk River (Ontario, Canada)	Pembroke residents	Drinking water	^{90}Sr	Bone	1‡	1963

† Based on continuous intake of river water since age 14 in 1944.
‡ Includes major contribution from fallout ^{90}Sr.

From Preston, A., 1969. In *Environmental Contamination by Radioactive Materials*, IAEA Symposium 117, p. 309; Crown copyright, reproduced by permission.

total local population consisted of a million persons. The theoretical acceptable value thus obtained was 1·3 mSv (0·13 rem) whereas the dose resulting from laverbread consumption was estimated to be 1 μSv (0·1 mrem).

Such are the vagaries of matters environmental that the laverbread critical pathway relating to the Windscale discharges does not apply at present; first of all the local railway staffing was altered, which affected the transport of *Porphyra* to market, and finally the two remaining collectors retired! Thus although still potentially a real critical pathway, and continually monitored because of this potential, the most important critical path for internal exposure at Windscale has more recently become the consumption of fish. Fish contain very little ^{106}Ru but do accumulate ^{134}Cs and ^{137}Cs. For such an important source of radioactivity as Windscale, a continual environmental surveillance is made of a large number of biological and non-biological materials, so that all potential pathways back to man can be evaluated. There is, in fact, one group of people in the Windscale area for whom external exposure is the critical path. These are salmon fishermen who work on mud flats attending their nets. Measurements are made of the dose-rate both on the nets and over the mud flats at monthly intervals and the mean dose-rates, coupled with a knowledge of time spent in the area, are compared with ICRP recommendations on acceptable limits. The gamma dose-rate is the result of ^{95}Zr/^{95}Nb and ^{106}Ru adsorbed onto the sediments.

The critical pathway approach has been used in a number of other countries. Some examples of different critical groups, materials, radionuclides and organs are given in table 5.7. The amounts of radionuclides authorized for discharge, and actually discharged, differ considerably from one nuclear site to another. The data in table 5.8 show that the total amount of radioactivity authorized for discharge at different sites may vary by four orders of magnitude, and that the actual amounts discharged in any one year may also vary considerably. The most important consideration, however, is the radiation exposure to the public resulting from these discharges. These estimates, for the discharge rates in table 5.8, are given in table 5.9. The Windscale site is a rather exceptional case, but it can be seen that although over 10 000 TBq (270 kCi) were discharged in 1975 the maximum exposure to a member of the critical group -- 34% of the ICRP-recommended dose limit – was only four times the value estimated for individuals around Lake Trawsfynydd into which less than 5 TBq (0·14 kCi) were discharged during the same period. Similarly, although Hinkley Point discharged about 7·4 TBq (0·2 kCi), and Sizewell only

TABLE 5.8. *Discharges to surface and coastal waters from Windscale – a reprocessing plant – and from some United Kingdom nucler power stations in 1975.*

Site	Radioactivity	Authorized discharge TBq (kCi) y⁻¹		Actual discharge TBq (kCi)		Actual discharge as % of authorization
Windscale	Total β	11100	(300)	9070·55	(245·15)	82
	^{106}Ru	2220	(60)	760·72	(20·56)	34
	^{90}Sr	1110	(30)	467·68	(12·64)	42
	Total α	222	(6)	85·1	(2·3)	38
Bradwell	Total activity†	7·4	(0·2)	4·403	(0·119)	60
	^{65}Zn	0·185	(0·005)	0·0037	(0·0001)	2
	^{3}H	55·5	(1·5)	3·293	(0·089)	6
Trawsfynydd	Total activity†	1·48	(0·04)	0·666	(0·018)	45
	^{3}H	74	(2)	3·33	(0·09)	5
	^{137}Cs	0·26	(0·007)	0·1776	(0·0048)	68
Hinkley Point	Total activity†	7·4	(0·2)	5·883	(0·159)	80
	^{3}H	74	(2·0)	1·961	(0·053)	3
Sizewell	Total activity†	7·4	(0·2)	0·74	(0·02)	10
	^{3}H	111	(3·0)	1·813	(0·049)	2
Wylfa	Total activity†	2·41	(0·065)	0·1295	(0·0035)	5
	^{3}H	148	(4·0)	4·773	(0·129)	3

† Excluding ^{3}H.

From Mitchell, N. T., 1977, *Radioactivity in Surface and Coastal Waters of the British Isles 1975*, Ministry of Agriculture, Fisheries and Food FRL 12; Crown copyright, reproduced by permission.

TABLE 5.9. *Estimates of public radiation exposure from liquid radioactive waste disposal from Windscale – a reprocessing plant – and from some United Kingdom power stations in 1975.*

Site	Exposure pathway to man	Maximum exposure of an individual, expressed as % of ICRP-recommended dose limit
Windscale	Fish consumption	34†
	Laverbread consumption	<0·2†
	External dose	9
Bradwell	Oyster consumption	0·07
Trawsfynydd	Fish consumption	8
Hinkley Point	Fish and shellfish consumption	<0·1
	External dose	<0·1
Sizewell	Fish and shellfish consumption	<0·1
	External dose	<0·1
Wylfa	Fish and shellfish consumption	<0·1
	External dose	<0·1

† To critical group.

From Mitchell, N. T., 1977, *Radioactivity in Surface and Coastal Waters of the British Isles 1975*, Ministry of Agriculture, Fisheries and Food FRL 12; Crown copyright, reproduced by permission.

2·6 TBq (0·07) kCi, the maximum exposures to individuals of similar critical pathways at the two sites were both less than 0·1% of ICRP-recommended dose limits. Even these few examples serve to illustrate the importance of careful evaluation of different sites, and also serve to illustrate the different capacities of sites to receive radioactive waste without undue exposure to man. This capacity is not entirely due to environmental factors, although these do play a part, but also to the differing habits of the local populations. Physical factors which are important are those of the volume and turn-over time of the receiving water, and the nature of the coastline in general. We shall return to these aspects in the next chapter. The human factors can be extremely variable, so that it is imperative to have a thorough and up-to-date knowledge of any change, or potential change, in the critical pathway. For this reason alone the critical pathway approach has particular merit, because the authorizing departments must maintain a close contact with that section of the general public most likely to be exposed to the discharged radioactivity. From the scientific point of view, perhaps one of the most surprising features is that the basic ecological assumptions, relating concentrations in different materials to a unit

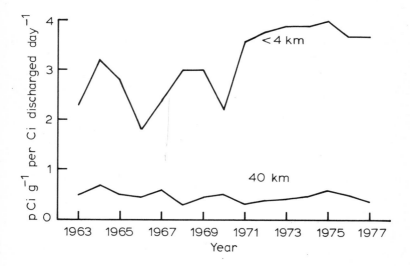

Figure 5.5. The concentration of [106]Ru in *Porphyra* collected at <4 km and at 40 km from Windscale for the period 1963 to 1977, normalized to a constant discharge rate of 1 Ci per day. (The ordinate can also be read as pBq g[-1] per Bq day[-1]). From unpublished MAFF material; Crown copyright, published by permission.

discharge, work so well. The concentration of [106]Ru in *Porphyra* normalized to a constant rate of input into the Irish Sea is shown in figure 5.5.

5.6 Solid wastes

During the life-time of a nuclear power station a considerable quantity of radioactive solid waste is expected to accrue. At magnox reactor sites, for example, about 2400 m^3 is expected to accumulate over a period of 25 years. About half of this volume, containing the majority of the 3 PBq (80 kCi) of radioactivity, will consist of an assortment of metallic components which have been removed from the reactor – control rods, chains, neutron flux measuring instruments, fuel stringer materials and so on. All of these materials will have been rendered intensely radioactive as a result of the absorption of neutrons by the stable ^{59}Co in the steel, producing ^{60}Co. This radionuclide is a high-energy gamma-emitter but fortunately has the relatively short half-life of 5·3 years. Because of the potential danger to personnel on site, however, it is necessary to store this solid waste in special concrete storage vaults within the biological shield of the reactor. The radioactivity will decay away to acceptable levels within a few decades, but as it is continually arising throughout the reactor's life it poses special problems for reactor decommissioning.

At the other extreme is a variable quantity of very low level waste. Sites such as reprocessing plants give rise to quite a large volume of low-activity waste and these may be buried at specially licensed disposal grounds. The Windscale plant disposes of some 3·7 TBq (100 Ci) of total activity in a volume of about 8000 m³ by this method every year. A third category of solid radioactive waste, which poses a more difficult problem of safe disposal, is that which contains very long-lived radionuclides. Some of this solid waste is packaged into drums and dumped into the deep ocean. This seemingly drastic action is very carefully controlled and undertaken under strict supervision.

At present those countries operating sea dumping do so within the terms of an international agreement called the London Dumping Convention, and operate to recommendations of the IAEA with regard to the procedures to be followed. The packaged waste conforms to recommended IAEA standards with regard to the concentrations of various classes of radionuclides – for example they must contain no more than 37 GBq t⁻¹ (1 Ci t⁻¹) alpha emitters (except ^{226}Ra), and no more than 37 PBq t⁻¹ (1 MCi t⁻¹) tritium. The IAEA recommendations also cover the conditioning and packaging of the waste, the selection of the dumping site, the method of dumping to be used, and the total amount to be dumped at any one site per year.

The packages are frequently, but not invariably, encapsulated in concrete, and are often contained within mild steel containers. The areas chosen for dumping are selected bearing in mind the chances of the packages being recovered by man – by bottom trawling or other forms of fishing, laying cables and so on; or the chances of the packages being moved out of the area by natural processes. Complex calculations are also made to ensure that, even if the contained radioactivity is released into the sea, the dose likely to be received by man, via any route, will be negligible.

A number of European countries combine their sea-dumping opera-tions, under the auspices of the Nuclear Energy Agency (NEA) of the Organization for Economic Co-operation and Development (OECD), currently dumping in an area some 900 km SSW of Land's End at an average depth of 4000 m. The NEA, in fact, provides a forum through which consultation can take place between interested parties – including countries which are opposed to sea dumping and wish to see what is going on. It also provides an independent observer of the dumping operations.

Convenient as this method may seem it is nevertheless not used at present by all countries. The USA, which in the past practised sea dumping, has more recently opted for land burial of solid waste instead, providing that it does not contain significant amounts of transuranium nuclides. These nuclides, of atomic number greater than that of uranium, must be segregated and stored such that they can subsequently be retrieved. The safe and economic disposal of solid waste is likely to be an increasing problem.

5.7. Collective dose to the public

In view of the multiplicity of potential pathways of exposure to man, and to different groups of the population, it is obviously very difficult to generalize on the overall impact on man's collective dose. Never-theless one faces the same difficulties when trying to generalize on the collective dose to man from the natural background, from radioactive fallout and from the medical uses of radiation. The difficulties should not, therefore, deter one from attempting to put public exposure to the routine releases of radioactivity from the nuclear power industries into some sort of perspective. The UNSCEAR reports therefore publish up-to-date summaries of collective dose commitments to the public. Their data published in 1977 for reactor operation and fuel reprocessing are given in table 5.10. The results are normalized to a MW(e)y.

All of this seems a little impersonal and a more detailed example may be more informative. As stated above, the dominant pathway back to man in recent years as a result of the Windscale reprocessing plant's low-level discharges to the Irish Sea has been the consumption of fish which contain ^{134}Cs and ^{137}Cs. The maximum individual dose as a result of fish consumption for the year 1975 was estimated to be 1·7 mSv (0·17 rem), 34% of the ICRP-recommended dose limit, and the average consumer received an estimated 0·35 mSv (35 mrem), 7% of the ICRP-recommended dose limit. The caesium discharged

TABLE 5.10. Normalized collective dose commitments to the public due to nuclear power production. Data in man-Gy/MW(e) y (man-rad/MW(e) y).

Source	Gonads	Whole lung	Thyroid	Bone marrow	Bone lining cells
Reactor operation †					
atmospheric	0·00211 (0·211)	0·00211 (0·211)	0·00311 (0·311)	0·00212 (0·212)	0·00212 (0·212)
aquatic	0·00040 (0·040)	0·00040 (0·040)	0·00050 (0·050)	0·00040 (0·040)	0·00040 (0·040)
Fuel reprocessing					
atmospheric	0·00002 (0·002)	0·00005 (0·005)	0·00202 (0·202)	0·00013 (0·013)	0·00014 (0·014)
aquatic	0·00130 (0·130)	0·00130 (0·130)	0·00430 (0·430)	0·00240 (0·240)	0·00240 (0·240)
Transportation	0·00003 (0·003)	0·00003 (0·003)	0·00003 (0·003)	0·00003 (0·003)	0·00003 (0·003)
Global contribution	0·01090 (1·090)	0·01250 (1·250)	0·01590 (1·590)	0·03250 (3·250)	0·03250 (3·250)
Total	0·015 (1·5)	0·016 (1·6)	0·026 (2·6)	0·038 (3·8)	0·038 (3·8)

† Incomplete dose commitments

Data from the Report to the United Nations Scientific Committee on the Effects of Atomic Radiation, 1977, General Assembly document 32 Session, Supplement No.40 (A/32/40) (New York: United Nations).

from Windscale, however, contaminates the sea water around a large fraction of the United Kingdom and fish from these waters are eaten by the British public at large and by the populations of other European countries. Data are therefore compiled on the quantities of fish landed in these countries, the concentrations of caesium radionuclides contained in samples of them – data considerably supplemented by concentrations computed from a knowledge of the sea water concentrations – and the size of the populations involved. The results for 1975 are given in table 5.11. The mean *per caput* doses may be compared with the annual absorbed dose to these populations of about 1 mSv (100 mrem) from natural background.

TABLE 5.11. *Collective and mean per caput dose rates resulting from Windscale discharges of caesium and the fish consumption pathway in 1975.*

Population (and size)	Collective dose rate man-Sv y^{-1} (man-rem y^{-1})	Mean *per caput* dose rate μSv/person y^{-1} (μrem/person y^{-1})
United Kingdom ($5{\cdot}5 \times 10^7$)	$8{\cdot}3 \times 10^1$ ($8{\cdot}3 \times 10^3$)	$1{\cdot}5$ (150)
Other European countries ($1{\cdot}4 \times 10^8$)	$5{\cdot}7 \times 10^1$ ($5{\cdot}7 \times 10^3$)	$0{\cdot}4$ (40)

From Mitchell, N.T., 1977, *Radioactivity in Surface and Coastal Waters of the British Isles 1975*, Ministry of Agriculture, Fisheries and Food FRL 12; Crown copyright, reproduced by permission.

5.8. *Future trends*

As to be expected, the more recent recommendations of the ICRP (No. 26) will alter, to some extent, the description which has been given in this chapter of the methods used to regulate the exposure of the general public to radionuclides via environmental materials. There will be changes in terminology: for example it is recommended that *derived limits* be set in defined models which relate either directly to dose limits or to secondary standards. More important, however, is the application of the change in ICRP philosophy. There are already differences in the application of ICRP guidelines at a national level. Having adopted the ICRP dose equivalent limits as the appropriate standards, some countries, such as the United Kingdom, have then made decisions on what is 'as low as reasonably achievable' for individual sites. Other countries, whilst again accepting the ICRP recommendations, have published 'standards' which, although specific to a particular situation, incorporate a general judgement on what is 'as low as reasonably achievable'. There are also national differences resulting from the methods of legislation. For example, in the USA the Environmental Protection Agency (EPA) issues, at a federal level, generalized 'standards' for most of the different stages in the nuclear fuel cycle; and the Nuclear Regulatory Commission (NRC), which actually licenses the nuclear installations, issues guidelines on such matters as radioactive effluents that are specific for individual reactors. These overall guidelines may then be further modified locally by authorities which are involved at the State, rather than the Federal, level of government.

Whichever method is adopted, however, it seems likely that the approach to environmental discharge assessments will become more sophisticated, even though the methods which have been used in the past have been extremely effective. It is customary, when considering discharges to the environment, to assess the pathways of different radionuclides back to man by taking into account the physical, chemical and biological characteristics of each particular site. The pathway analysis can then be used in two ways. In the first instance, assumptions are made which tend to maximize the dose estimate and the calculated estimates are then compared with the dose limits recommended by the ICRP. This approach, therefore, deliberately attempts to overestimate the dose to man, in order to be on the safe side. When considering the ICRP philosophy of optimization, however, it will be necessary to estimate the most probable distribution of doses which is likely to result from a particular method of waste disposal in order to calculate the collective dose, or collective dose commitment, as used in figure 5.1.

A more sophisticated approach is also to be expected in the development of mathematical models used in pathway analysis. In order to implement such control procedures as the critical path approach, particularly for radionuclides which do not have very long half-lives, it is not necessary to understand fully every aspect of an individual radionuclide's behaviour; it is sufficient to know that, for a unit rate of discharge, under equilibrium conditions, a certain concentration or dose rate will be obtained in the critical material. Pathway analyses have usually been evaluated using a concentration factor method: the control of ^{65}Zn discharges in relation to an oyster fishery, as described in section 5.5, is an example of this type of approach. The concentration factor method is very convenient because of its arithmetic simplicity. It does have certain limitations, however, because it does not take into account all time-dependent parameters. In other words, it not only assumes that some form of steady state condition will obtain in the environment, but that the radioactive releases to it will be more or less continuous. These two conditions are indeed often met, and such models have been found to be particularly useful in predicting the longer term, average dose to individual members of the public. But quite clearly this approach is not so applicable to situations involving non-uniform rates of radioactive discharge, and in such cases it is necessary to specify over what period of time a non-uniform rate of discharge can be averaged without invalidating the calculations based upon a uniform discharge rate. It is similarly difficult to predict the effect of acute releases, which may occur as a result of an accident.

An alternative, *systems analysis*, method is therefore becoming increasingly favoured. This method attempts to model the environment in a more dynamic way, using differential equations, and thus predict values along the pathway as a function of time, regardless of the type of release. It is an approach which requires considerably more data, and it also implies a much greater understanding of the individual components of a particular pathway. It is clearly more applicable to describing the eventual fate of the longer lived radionuclides, in as much as this can be described at all.

Mathematical models are sometimes regarded as being no more than substitutes for real data. Models used in pathway assessment, in fact, often help to identify those areas where data are particularly required, and thus the direction in which the research needs to be guided. This can be achieved by applying sensitivity analysis techniques. Tests of 'robustness' can also be applied; these attempt to quantify the extent to which the predicted doses to man are vulnerable to the combined effects of all the imprecisions inherent in the model.

The quantities and types of radionuclides deliberately released into the environment are dependent upon the engineering and plant operation procedures at the different stages of the nuclear fuel cycle, and upon the policies at national and regional level. The present trend is one in which the quantity of radioactivity released to the environment, per unit of electricity produced, is actually decreasing; but the number of sources is continually increasing. The total quantities of the longer-lived radionuclides which require ultimate disposal are also increasing. It is therefore quite evident that a greater emphasis will be given in the future to studies on the movements of radionuclides throughout the environment as a whole.

6. Environmental behaviour and effects

6.1. *Introduction*

So far the environment has only been discussed in relation to providing pathways of radiation exposure to man. And it is true that the major effort has been expended on ensuring that the quantities of radioactivity released into the environment are kept low enough for there to be no detectable detriment to man. There are, however, sound reasons for obtaining as full an understanding as possible of the behaviour, and effects, of radioactivity in the environment. First of all there are philosophical reasons: if potentially noxious substances are to be deliberately discharged into the environment then one should know what happens to them. Secondly, it is necessary to learn something of the mechanisms by which radionuclides are transported under different environmental conditions in order to make predictions for new sites, to make predictions of the long-term fate of the longer lived radionuclides, and to employ mathematical models such as those of systems analysis. Thirdly, it is obviously necessary to consider the effects of radiation on the different components of the ecosystem; not only because of the possible deleterious effects on the fauna and flora in general – a perfectly laudable and sufficient reason in itself – but because man may also be disadvantaged if important food resources are adversely affected. There is no equivalent to the set of ICRP guidelines – which relate only to man – for any other species. Because of all these reasons a considerable research effort has been mounted over the years to learn a good deal about the environmental behaviour and effects of radioactivity. All sources of information have been used – naturally occurring radionuclides, fallout from the atmospheric testing of nuclear weapons, and the controlled authorized discharges of low-level waste from the nuclear industries. The study of radioactivity in the environment has even become a recognized branch of science in its own right– radioecology.

It is difficult to summarize briefly the large amount of data

172

available, although it is again useful to recall that a number of the radionuclides released into the environment are isotopes of elements essential to life. The list includes not only carbon, hydrogen, oxygen, nitrogen and sulphur, which form the basic building units of proteins, carbohydrates and lipids, but also the essential trace elements. Amongst the latter are a number of metals which, either as ions or in combination act as catalysts, such as iron, manganese, cobalt and zinc; and elements which are essential structurally, such as barium, iron and phosphorus. It will also be recalled that other elements, although inessential in themselves, are chemically similar to elements which are essential; for example caesium is similar to potassium, and strontium to calcium. Radioisotopes of all of these elements are, or have been, released into the environment.

A number of radionuclides are, indeed, isotopes of elements which have no known biological function – for example cerium, ruthenium and zirconium. In addition there are the transuranium elements, which are essentially man-made. The degree to which both animals and plants assimilate and metabolize these elements frequently reflects such differences, although many non-essential elements are accumulated, often by direct physico-chemical processes, to a remarkable degree.

6.2. Behaviour in terrestrial ecosystems

The terrestrial environment may become contaminated in a number of different ways; by the direct deposition of contaminated soil, by flooding – including irrigation practices – with contaminated water, or by aerial deposition, both wet and dry. Terrestrial plants may, therefore, either become contaminated as a result of direct deposition, or indirectly as a result of radionuclides first entering the soil, or both. There are two principal factors which affect direct contamination; the physical characteristics of the aerial deposit and the growth form of the plant. Radionuclides may be either *ad*sorbed or *ab*sorbed by the above-ground parts of the plant, although of course terrestrial plants can only absorb elements if they are in solution. Particulate deposits having a diameter greater than about 50 μm are not normally retained on leaves; smaller particles, however, about 5 μm and less, will collect directly onto the surface of the leaves. The general morphology of the plant is also very important; tufted species, such as many of the grasses, being very efficient funnel collectors of particulate radio-activity. For very fine particles, such as those likely to occur in ^{131}I deposits, deposition rates will be affected by laminar boundary diffusion processes. The laminar boundaries are non-turbulent layers of air which surround the surface of the leaf: the thicker the laminar boundary layer the lower the deposition rate. The boundary layers are disturbed by protuberances on the leaf's surface and are disrupted by turbulence at the leaf's edge. The deposition of fine particles thus occurs irregularly on the surface of leaves, and along the edges.

The leaves, of course, are not the only above-ground part of a plant; floral parts – inflorescences – can also be important, particularly for cereals. It has been observed that the heads of crops such as wheat are very efficient collectors of fallout particles; and cereals have been a major source of ^{90}Sr in the diet of man as a result of the atmospheric testing of nuclear weapons. In fact, white bread has been found to contain less ^{90}Sr than brown bread because the latter includes the test of the grain, to which the ^{90}Sr is adsorbed. This fact also emphasizes the importance of the season on the degree to which a plant, and thus the food chain, is contaminated. The importance of the season is also demonstrated by a third source of above-ground contamination, the plant base. Radionuclides adhering to the foliage can be removed not only by radioactive decay but by straightforward weathering effects – leaching by rain, processes of volatilization, mechanical effects – and by the withering of the foliage itself. These processes will result in the accumulation of radionuclides around the basal structures of plants, particularly of perennials which develop a ground level mat of old stems and surface roots. Such accumulation and concentration of longer lived radionuclides may result in absorption through the basal parts of the plant, or through surface roots, in the subsequent growing season. This can be an important source of contamination for pasture-land.

It is usually assumed that the foliage parts of a plant have a minimal capacity to absorb chemicals other than oxygen and carbon dioxide; an assumption which is readily shown to be false by the widespread application of herbicides which are rendered innocuous upon contact with the soil. Thus, although the major fraction of directly contaminating radioactivity does remain on the surface of the foliage, and will subsequently be removed, some absorption does take place. Iodine is one element known to be absorbed, although for the principal isotope of concern in accident situations, ^{131}I, the short half-life of 8 days minimizes the importance of any subsequent translocation within the plant. Caesium, on the other hand, is absorbed via the basal portions of plants and is readily translocated, resulting in a fairly uniform distribution; with ^{134}Cs and ^{137}Cs having half-lives of 2 and 30 years, respectively, this has important consequences for both root and non-root crops. A third element of particular interest is strontium, which although entrapped by inflorescences of crops such as cereals, and indeed any part of a plant with a relatively large surface area, is hardly absorbed at all at these sites. There is also very little downwards translocation. Thus leafy brassicas, such as cabbages, do not reflect ^{90}Sr fallout in the same manner as cereal crops, when processed for human food, because the outer leaves are removed and not eaten. Similarly, root crops do not immediately reflect ^{90}Sr contamination although ^{90}Sr will be accumulated subsequent to any aerial deposition, both through the basal portion of plants and via the principal route of entry for chemicals into plants – the soil.

Soils are very complex mixtures of inorgânic and organic matter which also contain water and dissolved gases; their physical and chemical properties vary enormously. Soil is vertically stratified into zones normally referred to as surface soil, subsoil – the two collectively termed the regolith – and parent material or bedrock. The texture of the soil depends upon both the nature and content of the organic and inorganic fractions. The latter will, in turn, be governed by the percentage of sand, silt or clay. These three classes are largely differentiated by size. Sand particles vary from about 2 mm down to ~50 μm diameter, silt from ~50 μm down to ~4μm, and clay particles are less than ~4 μm diameter. Between these three basis classes a wide range of soils can be characterized, ranging from pure sand, silt or clay to a fairly evenly balanced mixture called loam. When one type predominates in a mixture the soil may be termed, for example, sandy-loam and so on. The organic content can also vary greatly so that a soil may be rich or poor in humus. The relevance of all this to radioecology is that the nature of the soil, both physical and chemical, has a great effect on the concentrations of different radionuclides which it can retain, and the extent to which such retained radioactivity will be available for absorption by plants.

The presence of clays is particularly important because it is this fraction which provides the chemical reservoir of the soil. Very little is held at any one time by the soil water – it would be rapidly washed away. The surfaces of clay particles are excellent ion exchangers. They have a particular affinity for positive ions and the exchange of these with the water in the soil usually leads to the soil gradually becoming acidic. This is because adsorbed cations are slowly replaced by hydrogen ions. Soils are 'limed' for this reason – replacement of the hydrogen ions by those of calcium and magnesium. In view of this ion exchange capacity it is to be expected that divalent cations would be more strongly bound than monovalent cations; but relative chemical abundance is also important. It may at first seem surprising that a chemical such as caesium may be more tightly bound to soil particles than strontium, but the latter is more abundant. There is also competition by similar ions, however, so that the retention of both caesium and strontium will be affected by the presence of potassium and calcium, respectively. Soils having a high calcium content exhibit a similar rate of penetration of ^{90}Sr and ^{137}Cs whereas in more acidic soils the former nuclide is relatively more mobile and more quickly moves downwards into the soil matrix.

All of these factors will, in part, affect the degree of penetration of the radionuclide into the soil, an important consideration in terms of subsequent availability within the biosphere. Such is the extent of adsorption, however, that the majority of studies have demonstrated how slowly most radionuclides of interest migrate downwards. For example, analyses made in 1974 of a flood-plain forest ecosystem in the Tennessee valley, in the USA, which was contaminated with ^{137}Cs in 1944, revealed that practically all of the radionuclide was still within 60 cm of the soil surface; maximum concentrations were within a band some 12 to 22 cm from the surface. The presence of leaf-litter on the surface also has an important effect on the degree of penetration into the soil; but the overriding factors, in farmland, are usually the degree of irrigation, topography and agricultural practice. A rather extreme example of this last factor is that of an experiment made in Washington State in the USA where ^{137}Cs and ^{90}Sr were applied to plots of bare soil, some of which were left untilled and others tilled with a variety of crops for 8 years. In the untilled plot, which maintained a continual population of annual weeds, 70% of the ^{137}Cs remained in the upper 3 cm of the soil whereas in the tilled soils the ^{137}Cs was mixed to a depth of 13 to 16 cm.

There have been few studies on the role of the soil fauna in the distribution of radionuclides with soils, but it is known that burrowing animals such as earthworms do play a part in redistributing radionuclides. In a deliberately labelled forest of tulip-trees (*Liriodendron tulipifera*) the dominant earthworm (*Octolasium lacteum*) was estimated to have turned over about 15% of ^{137}Cs present in the soil over a period of 7 years.

It is thus extremely difficult to generalize on the behaviour of radionuclides in soils, or on any subsequent accumulations by plants. Approximations have, of course, been made. The data in table 6.1 have been derived by determining the extent to which a number of elements, added to the soil in a water-soluble form, have been enhanced by crops grown in it. Some ions exist naturally in solution in soil water

TABLE 6.1. *Relative concentration of elements in the first-crop plants grown after the elements have been applied in water-soluble form and mixed into surface soil. The relative concentration factors (RCF) have been expressed as* $\mu g\ g^{-1}$ *dry plant material divided by* $\mu g\ g^{-1}$ *dry soil.*

RCF Range				
10 – 1000	1 – 100	0·1 – 10	0·01 – 1·0	< 0·01
K	Mg	Ba	Cs	Sc
Rb	Ca	Ra	Be	Y
N	Sr	Si	Fe	Zr
P	B	F	Ru	Ta
S	Se	I	Sb	U
Cl	Te	Co	W	Ce
Br	Mn	Ni	Hg	Pm
Na	Zn	Cu	As	Pb
Li	Mo		Cr	Pu
			V	Am
				Cm

From Menzel, R.G., 1977. In *Proceedings of a National Conference on Disposal of Residues on Land*, (Rockville, Maryland: Information Transfer), p.93; courtesy of the author.

but the total, labile, ionic pool also includes those which are bound exchangeably to both organic, and particularly the clay fraction of the inorganic, soil components. Typical of this labile ionic pool are calcium, potassium, chloride, nitrate and sulphate. The abundance of calcium in most productive soils is such that its presence considerably reduces the accumulation of ^{90}Sr by plants, even in acidic soils where it is more mobile. The considerable effort and expense of liming soils to reduce ^{90}Sr uptake by plants is thus only really worth considering for soils which have a low-productivity potential in any case.

Leaf litter can provide a continual route of input into the local ecosystem, in contrast to agricultural land, where radionuclides accumulated by plant materials are cropped. The direct inoculation of trees via the xylem has provided a variety of data on the relative distribution of radionuclides throughout such large plants and the effect this has on leaf litter content. Flowering dogwood (*Cornus florida*) injected with ^{45}Ca at the base of the stem were observed to transfer a maximum of 73% of the nuclide to the leaves within one month. The fate of this 73% was that 64% entered the leaf litter by leaf fall, 6% was leached off the intact leaves on the tree, and the remaining 3% was accounted for by insect damage to the leaves. In another experiment, in which red maple (*Acer rubrum*) were injected with ^{45}Ca, the subsequent leaf litter was collected and laid on the floor of an oak-hickory forest. Within 18 months half of the ^{45}Ca had moved into the soil.

It is believed that the behaviour of strontium would be similar to that of calcium; caesium, however, is known to behave quite differently. Tulip-trees injected with ^{137}Cs have been shown to lose only 8% via litter fall and leaching from the leaves, whereas 45% was accounted for by a combined soil, root and leaf-litter mixture. Disentangling this mixture was not entirely successful, but it did appear that approximately a third, that is 15% of the total injected, was that of the soil plus root component of the budget. It has not been shown conclusively, however, whether this is a result of ^{137}Cs transfer to the soil by live roots or the result of root death and decay, which is also a continuing process. Regardless of the method of transfer the ^{137}Cs is then evidently bound to the soil, especially those soils having a high clay content. An interesting difference in the behaviour of caesium has been observed in the state of Florida where the fauna and flora generally have been noted to contain more ^{137}Cs than other areas in the USA which have received a similar amount of fallout. The soil in Florida is sandy and generally of low fertility. Attempts to leach ^{137}Cs from a variety of soils did not indicate that enhanced uptake by plants had resulted from increased leaching into soil water, and it has been surmised that one factor influencing plant uptake is the direct recycling of ^{137}Cs from the leaf litter to the root system via mycorrhiza. This is a rather intriguing pathway and one deserving of greater study.

It was stated in chapter 5 that terrestrial food chains ending in man are usually fairly short, such as that from grass to cow, and from cow and its dairy products, to man. Even in this short food chain a considerable exclusion of radionuclides may be observed, as demonstrated in a study made of the milk produced by a 'family cow' shortly after the atmospheric testing of a nuclear weapon by the Chinese in 1966. Many small farms in the USA have a 'family cow' and this particular one was an 8-year-old Hereford-Angus. As a result of dry deposition alone, the grass upon which she grazed was found to contain a wide range of gamma-emitting radionuclides. In addition to the ubiquitous ^{40}K, the following were also apparently found: ^{95}Zr/^{95}Nb, ^{99}Mo, ^{103}Ru, ^{125}Sb, ^{131}I, ^{132}Te/^{132}I, ^{133}Xe, ^{137}Cs, ^{140}Ba/^{140}La, ^{141}Ce, ^{147}Nd, ^{238}Np and ^{239}Np. Of these, again apart from ^{40}K, only ^{131}I, ^{137}Cs, ^{99}Mo, ^{140}Ba/^{140}La and ^{132}Te/^{132}I were detectable in the milk. Such fallout studies as this, and numerous other studies on the metabolism of elements by vertebrates – particularly mammals – using both radionuclides and stable elements, have provided a wealth of data on the transfer of nuclides through a number of important food chains. The extent to which any element is accumulated represents a balance between the rate at which it is taken in and the rate at which it is excreted. By and large, univalent cations diffuse readily across the vertebrate gut wall and the majority are excreted via the urine. Less readily absorbed are divalent cations, some of which are excreted via the urine and others excreted by being transferred back into the gut lumen and eliminated via the faeces. The excretion of divalent cations is usually slow. Polyvalent cations are not readily absorbed. In contrast anions, with some exceptions, are generally well absorbed. The subject is obviously a very complicated one and to say 'generally' is to be almost flippant. A large number of data have arisen from studies on animal nutrition and these have highlighted the importance of a knowledge of the total diet on any estimations of elemental assimilation. Thus it is known, for example, that the absorption of barium and molybdenum is affected by the presence of sulphate, that calcium absorption is affected by vitamin D, and so on.

Terrestrial invertebrates may either have permeable skins, which restrict them to damp habitats, or have relatively impermeable skins that reduce water loss. The latter are typified by those most successful colonizers of land, the insects and arachnids, whose ability to conserve water is such that elimination of nitrogenous waste may be effected by the excretion of crystals of uric acid. Even permeable invertebrates such as the annelids are unlikely to absorb many nuclides directly through the skin, which is covered in any case by a cuticle that to some extent reduces water loss. Earthworms, it appears, obtain radionuclides such as ^{137}Cs by ingestion of contaminated leaf litter.

Some of the more detailed studies on the transfer of radionuclides along terrestrial food chains have been made on forest arthropods. Nuclides such as ^{137}Cs have been shown to decrease in concentration from primary arthropod consumers – saprovores – to primary arthropod predators. Even the saprovores attain concentrations which are only about 70% of that of the detritus upon which they feed. Other radionuclides appear to differ; ^{106}Ru and ^{60}Co are both accumulated by herbivorous arthropods – grasshoppers, leafhoppers, weevils – more than ^{137}Cs, and predaceous arthropods such as arachnids and coccinellids may attain ^{106}Ru concentrations approximately twice those of herbivores.

Terrestrial food chains of vertebrates have demonstrated the reverse of the above findings; in contrast to the arthropods, the vertebrate gut shows a very poor assimilation efficiency for [106]Ru. Caesium, however, is readily absorbed by both invertebrates and vertebrates, but the concentration of ^{137}Cs in vertebrate predators is usually greater than that of prey species because the former have longer biological half-times, a reflection of their much greater size. The data given in table 6.2 illustrate that carnivores such as the bobcat may contain, on a whole-body basis, some 6 to 16 times as much ^{137}Cs as its prey, depending upon geographical location.

TABLE 6.2.   *Ratios of predator/prey whole-body burdens of ^{137}Cs observed in the coastal plain area of South Carolina, and in the Georgian Piedmont areas of the USA.*

Species	Locality	Mean ratio
Bobcat/rabbit	Coastal plain, S.C.	15·9
Bobcat/rabbit	Coastal plain, S.C.	14·0
Gray fox/cotton rat	Coastal plain, S.C.	5·6
Bobcat/rabbit	Piedmont, Ga.	6·1
Gray fox/cotton rat	Piedmont, Ga.	2·0

From Jenkins, J.H., Monroe, J.R., and Golley, F.B., 1969, *Symposium on Radioecology*, Second National Symposium CONF-670503, p.623; courtesy of the authors.

The transfer of a number of other radionuclides along terrestrial food chains has been studied and, as may be expected, many variables have been found to be influential in the process. The feeding habits of animals at different trophic levels are primarily accountable for differences between vertebrates. Carnivorous animals which swallow their prey whole, such as foxes, or snakes, ingest not only the radio-nuclides contained within their prey but also radionuclides adhering to the skin, fur or feathers. Other carnivores are more discriminating: shrews will selectively eat certain soft tissues of their prey – brain, heart, liver, muscle – but will not ingest intestines or skeletal material. Different again are raptors such as owls and hawks which, although swallowing their prey whole or in large pieces, will regurgitate indigestible parts such as fur, skin and bone in the form of pellets.

No account of terrestrial food chains would be complete without mention of what has become a classic study, the transfer of ^{137}Cs from lichen via reindeer to some inhabitants of the Arctic. The ^{137}Cs originated from the atmospheric testing of nuclear weapons, and lichen proved to be particularly efficient at retaining a number of fallout radionuclides. This particular food chain attained a certain notoriety because a number of Laplanders and Eskimos attained ^{137}Cs whole-body burdens in the range of 37 kBq (1μCi); higher than other populations, although of no radiological significance.

Such food chains are of little relevance to the release of radionuclides from nuclear facilities, however, and indeed it may well be questioned whether the majority of data which have been discussed in this section are of any relevance at all to nuclear waste disposal. It is true that the majority of studies which have been made relate to fallout from the atmospheric testing of nuclear weapons, and from other studies on the longer lived fission product radionuclides. Such data are of value in predicting the extent of contamination resulting from accidental releases of radioactivity from nuclear facilities − or from nuclear war − but do not relate very much to the day-to-day releases of a modern reactor as discussed in chapters 4 and 5. In fact, when studies are made of vertebrate species around nuclear facilities, very few radionuclides are observed. Wild birds have been found to contain traces of ^{131}I in the vicinity of a boiling water reactor during shut down for refuelling, but in the same study the presence of other radionuclides, such as ^{137}Cs in muscle, were found not to correlate with any known routine operation at the site.

The largest quantities of radionuclides introduced into the atmosphere are those of the noble gases, and their impact on animals and plants is not expected to be any different from that on man. The behaviour of other radionuclides, for example those of carbon and sulphur, may be generally inferred from the basic knowledge of their cycling in the biosphere and measurements made of their specific activities. The carbon cycle has been particularly well documented. It begins with the fixation of carbon dioxide from the atmosphere by plants via the process of photosynthesis; carbon dioxide and water react to form carbohydrates, with the simultaneous release of free oxygen. Some of the carbohydrate is used by the plant as a source of energy, releasing carbon dioxide back into the atmosphere; and some is consumed by animals which, in turn, release carbon dioxide back into the atmosphere, both as a result of their respiration and by their death and subsequent decay. This discrete cycling process looks very neat and tidy but in fact the actual amounts involved, the relationship between this cycle on land and a similar but more complicated one in the seas, and the ultimate fate of carbon which becomes stored in sedimentary rocks − all are areas of research which are the subject of much controversy.

Sulphur is another essential constituent of living organisms; it is the bonding between sulphur atoms which maintains the three-dimensional shapes of proteins. Sulphur is also cycled in the biosphere. The general picture has been that sulphide in the atmosphere is oxidized to sulphur dioxide which, after dissolution in rainwater and further oxidation, enters the biosphere as sulphate. Sulphate-reducing bacteria complete the cycle. More recent research has turned towards evaluating the role played by the oceans in the cycling of sulphur. In order to account for the quantities of sulphur dioxide observed in the atmosphere, it has been suggested that some of it may be derived from dimethyl sulphide produced by marine organisms. Dimethyl sulphide may well be released into the atmosphere and subsequently oxidized to sulphur dioxide. The ^{35}S nuclide has the relatively short half-life of 87 days, and the cycling of sulphur in the long term is, therefore, of little consequence, unlike that of ^{14}C. Radionuclides with half-lives far longer than ^{14}C are being increasingly introduced into the environment, however, and it is essential to learn more of their long-term behaviour. One radionuclide of immediate interest is ^{129}I which, with a half-life of 1.6×10^7 y, will remain in the environment for a very long time. The cycling of iodine in the environment is not clearly understood. The beta-emitting ^{99}Tc, with a half-life of 2.1×10^5 y, is another nuclide which needs to be studied in more detail. Increased attention is now being paid to the transuranium nuclides; fortunately it appears that these elements are very well adsorbed by soils, and that little transfer takes place from soils to plants. Plutonium which does become incorporated into plants is better absorbed across the gut than when it is in solution, however, and the degree to which such findings apply to the other transuranium nuclides needs studying. In fact it is to those elements which are more directly related to nuclear power wastes that attention should be paid in the future, the longer lived nuclides requiring particular attention, to assess their potential environmental distributions.

6.3. Behaviour in aquatic ecosystems

Unlike the majority of terrestrial radioecological studies on fallout, those made on the aquatic environment have been of direct relevance to the discharge of liquid wastes arising from the nuclear industries because they contain a similar range of nuclides. It is in many ways fortunate that a large number of atmospheric weapons tests were made in and over the sea, for it is the marine environment which receives the major fraction of the liquid waste discharges. The adjective 'fortunate' applies not merely to the useful data which were gained as a result of such studies, but to the enormous stimulus such events gave to research into the subject of marine chemistry, particularly with regard to trace elements. The sea contains all of the elements. Apart from hydrogen and oxygen, in the form of water, dominant among them are chlorine at 19 g l^{-1} and sodium at 11 g l^{-1}. Also abundant are magnesium (1·3 g l^{-1}), sulphur (0·9 g l^{-1}), calcium (0·41 g l^{-1}) and potassium (0·39 g l^{-1}). Carbon, both inorganic and organic, bromine and gaseous nitrogen are all present at concentrations in excess of 10 mg l^{-1}. All of these elements are referred to as being 'conservative' to sea water because their concentrations are directly proportional to the salinity. Other conservative elements which are of particular interest to the radioecologist are strontium, caesium, uranium, iodine, and the inert gases. Many elements which are not conservative are those such as nitrogen and phosphorus – in the form of nitrate and phosphate, respectively – and iron, which are involved in biological cycles and therefore vary both in geographical distribution and with season.

The sea is, therefore, by no means homogeneous with regard to chemical composition, and coastal waters are less homogeneous than the open ocean. There are a number of reasons for this difference, principal among them being the rapid mixing of substances introduced via rivers, circulation patterns which favour the retention of dissolved chemicals in the waters near the coast, the relative abundance of particulate matter, and the very much higher biological activity. In the majority of coastal waters, precipitation and freshwater run off exceed evaporation from the sea's surface. The less dense freshwater mixes in the surface layers and moves away from the coast, being replaced by a sub-surface input of more saline water from offshore. This process also helps to maintain the chemical inhomogeneity by returning to the coastal waters those chemicals which, by a variety of processes, have settled out from the surface layers. Indeed it has usually been considered that the returning of certain chemicals by this process to the surface layers, to which light can penetrate, aids the large primary productivity so characteristic of shelf seas.

It follows, therefore, that the degree of dispersion of radionuclides released into coastal waters will depend upon the physical characteristics of the receiving water mass and the chemical properties of the element. In view of the relatively greater quantities of radio-activity discharged by reprocessing plants, the distribution of ^{137}Cs and [106]Ru discharged under authorization by Windscale into the Irish Sea will serve as an example of how disparate in behaviour two radio-nuclides can be. The lines of equal concentration of the two radio-

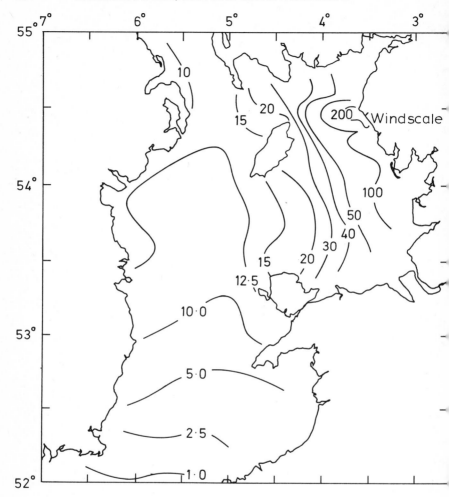

Figure 6.1. The contours of concentration (pCi l⁻¹) ^{137}Cs in filtered sea water from the Irish Sea in July 1973 (1 pCi = 37 mBq). From Hetherington, J. A., Jefferies, D. F., and Lovett, M. B., 1975. In *Impacts of Nuclear Releases into the Aquatic Environment*, IAEA Symposium 198, p.193; Crown copyright, reproduced by permission.

nuclides in filtered sea water shown in figures 6.1 and 6.2 indicate that the concentration of ^{137}Cs near the Isle of Man is a tenth that on the Cumbrian coast whereas that of ^{106}Ru is only a fiftieth. This is partly due to radioactive decay, but is also the result of a much greater adsorption of ^{106}Ru on to sediments and particulate matter in

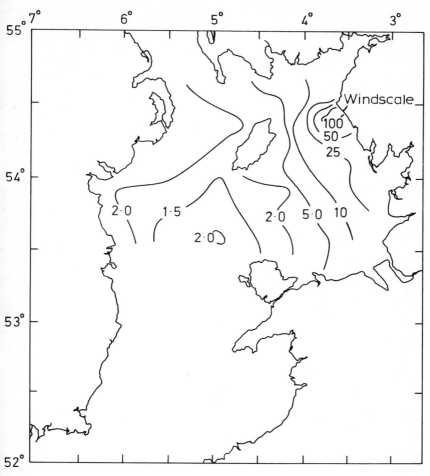

Figure 6.2. The contours of concentration (pCi l^{-1}) ^{106}Ru in filtered sea water from the Irish Sea in July 1973 (1 pCi = 37 mBq). From Hetherington, J. A., Jefferies, D. F., and Lovett, M. B., 1975. In *Impacts of Nuclear Releases into the Aquatic Environment*, IAEA Symposium 198, p.193; Crown copyright, reproduced by permission.

suspension. The conservative nature of caesium, and the fairly long half-lives of two of its radionuclides – 30 years for ^{137}Cs and 2 years for ^{134}Cs – has made it an extremely useful tracer of large-scale sea water movements. Thus it is possible to trace the movement of water out of the Irish Sea, around the west coast of Scotland, and down into the North Sea. If the ratio of the two nuclides in the discharges from

Windscale remains constant for reasonably long periods, which it usually does, estimates can be made of the transit times of the sea water by making use of their two different rates of isotopic decay.

Caesium is a monovalent element and has a relatively simple chemistry in sea water. Other elements have a far more complex chemistry, and this is reflected in the degree to which they are removed from sea water; it also implies that the chemical form of the radionuclide at the time of discharge may affect its subsequent behaviour. Radioisotopes of ruthenium discharged into sea water are known to exist in a variety of complexes. Little is known of their discrete behaviour, however, although it does appear that the nitrosyl-chloride form may predominate. Recent studies on plutonium in the Irish Sea have shown that this element behaves in different ways according to its chemical state. Sea water samples are usually sub-divided for analysis into filtrate and particulate by passing them, typically, through a $0.22 \mu m$ filter. When Irish Sea water is so treated, the plutonium which passes through the filter has been shown to be $Pu(V + VI)$ whereas that retained by the filter paper is $Pu(III + IV)$. The relevance of such differences in chemical speciation is reflected in the degree to which radionuclides are accumulated by the biotic and abiotic components of the environment. The chloride form of ruthenium, for example, is accumulated to a much greater extent by bivalve molluscs than is the nitrosyl-nitrato form; and sedimentary particles in suspension accumulate Pu $(III + IV)$ over the surrounding water by at least two orders of magnitude greater than $Pu(V + VI)$. The presence of other elements in the sea water also affects the degree of removal of radionuclides to the sea bed. The concentrations of both iron and manganese in sea water decrease markedly with increasing distance from the coast, iron more than manganese, and both are influential in co-precipitating other elements.

Non-conservative radionuclides, therefore, are fairly rapidly removed to the sea bed. Marine sediments display all the variables of terrestrial soil; they vary in texture from beds of gravel – inorganic and shell-gravel – through sand to fine mud. Many areas of the sea bed consist either of bare rock or hard clay. The organic fraction of marine sediments also varies greatly from one extreme to another. Radionuclides adsorb on to the surfaces of particles and it has generally been observed that, because this is a surface area phenomenon, the amount of radionuclide per unit weight of a sample is inversely proportional to the diameter of the particles over a fairly broad range (figure 6.3). The relationship does not necessarily hold for very fine particles because their surface areas may be greatly increased by a

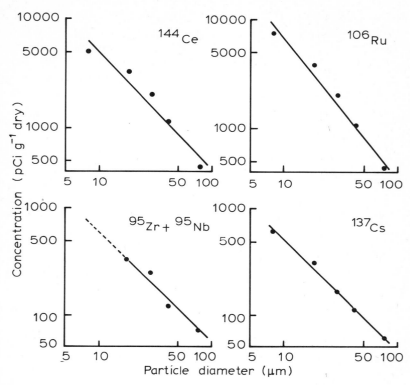

Figure 6.3. Relationship between the concentration of [144]Ce, [106]Ru, [95]Zr/[95]Nb and ^{137}Cs and particle diameter in silt and fine sand from the Windscale area. From Hetherington, J. A., and Jefferies, D. F., 1974, *Netherlands Journal of Sea Research*, **8**, 319; Crown copyright, reproduced by permission.

matrix within the fine clay materials. Mineral composition is also important, however, and different sized particles within a single sea bed sample frequently have a quite different mineral origin. Carbonate minerals are considered to have a lower capacity to adsorb a number of radionuclides owing to their lower cation exchange capacities.

The continued input of radionuclides into an area usually results in contamination of sediment in a vertical direction. This may be the result of more than one process. Freshly labelled particles may settle out on top of the existing particles, a process known as accretion. Radionuclides may also diffuse down the sediment column by way of the interstitial (or pore) water and thereby adsorb to particles lower in the column. Core samples of marine and estuarine sediments therefore frequently display profiles of concentration down the column which

usually relate to the varying degrees of input at the surface, and to the
rate of decay of the radionuclide. If the rate of input at the surface is
normalized, i.e. reduced to a unit value, it is possible to compute half-
depths – the depth at which the surface concentration is reduced to
one half – for different radionuclides. To take another example from
the well-labelled Irish Sea, figure 6.4 illustrates the exponential rate

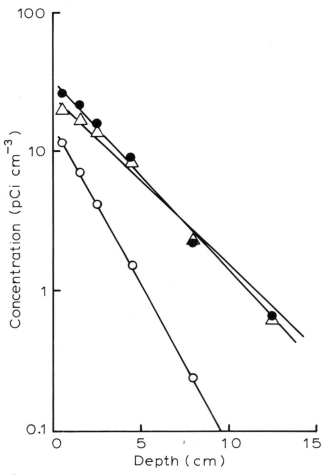

Figure 6.4.   The distribution of radionuclides with depth in the mud of an
estuary near Windscale, normalized to a discharge rate of 1 Ci^{-1} (1 Ci = 37
GBq and 1 pCi = 37 mBq). The nuclides are: ^{144}Ce (●), ^{106}Ru (Δ) and
^{95}Zr/^{95}Nb(○). From Hetherington, J. A., and Jefferies, D. F., 1974,
Netherlands Journal of Sea Research, **8**, 319; Crown copyright, reproduced by
permission.

of decline in the concentration of $^{95}Zr/^{95}Nb$, ^{106}Ru and ^{144}Ce in estuarine mud normalized to a discharge rate of 37 GBq (1 Ci) per day from Windscale.

Data such as these are used to make estimates of diffusion constants into the sediments. There are still a large number of unknowns in this area of research, however, and it is a field of continuing observation and experimentation. Of prime concern both for an understanding of the mechanisms involved, and for long-term predictions, is a knowledge of the permanency of the radionuclide-to-particle bond: such knowledge is often tainted by the methods used to obtain it – as is so much of research data. But it is known that the chemical environment may change considerably with only a short increase in depth, particularly in areas with a high organic content, producing a change from oxidizing to reducing conditions. The effects such changes may have on the chemical speciation of radionuclides entrapped in sediments is an important area of current research. A further complication is the presence of the benthic fauna, those animals which live both on and in the sea bed – the epi-fauna and in-fauna, respectively. They may be extremely abundant in coastal and estuarine waters, and their movements may have several effects. The top few centimetres of many sea bed cores are frequently well mixed with regard to the concentration of radionuclides within them, and this is undoubtedly often the result of the physical re-working of the sediment by benthic animals. Burrowing animals may also form vertical channels which could permit the exchange of sea water at the surface with that deeper in the sediment. The importance of such events, and the role of bacterial activity in the sediments, is largely unknown.

The interactions of radionuclides with water and sediments, and their chemical behaviour in these media, have a very marked effect on their accumulation by aquatic organisms. Radionuclides may be adsorbed or absorbed directly from the water and subsequently passed along a complex array of food chains. Aquatic food chains are often so complex that the term food web is used to emphasize their intricate diversity. At the base of all food chains are the plants, and it is the algae which predominate in the sea. Benthic algae – multi-cellular algae attached to the sea bed – occur on suitable coasts from the inter-tidal zone down to a depth of several metres, depending upon the turbidity of the water and thus the penetration of light. Both plants and animals which colonize such habitats are liable to severe mechanical damage by wave action. Benthic algae are coated in mucus, of which there a variety of chemical forms which differ from species to species. This external covering provides a slippery surface that minimizes wear and tear, and also helps to minimize water loss should the plant be exposed at low-water. But, of particular value to the radioecologist, it also provides an excellent surface for the adsorption of a large range of radionuclides. The concentrations of radionuclides on algae are, therefore, frequently far in excess of those of the surrounding sea water. This property, together with their sessile nature, makes benthic algae very useful indicators of the presence of certain radionuclides whose lower concentrations in sea water would otherwise make them very much more difficult to detect. The accumulation of [106]Ru by one specific alga, *Porphyra umbilicalis*, has been the subject of detailed study because for many years, as noted in chapter 5, it was the critical material related to the authorized discharge of low-level radioactive wastes from Windscale. Both living and dead discs of *Porphyra* tissue accumulate [106]Ru from sea water at a similar rate. The accumulation has been shown to be pH dependent – being suppressed at lower pH values – and it has therefore been concluded that cation exchange reactions are involved at the algal surface. Experiments with other algae have largely substantiated the role of surface adsorption in radionuclide accumulation by benthic algae. In particular, the accumulation of zinc by different species has been shown to be the result of non-metabolic processes. It has also been shown that different specific activities can be attained by the large brown seaweed, *Laminaria*, in sea water to which both a fixed quantity of [65]Zn and a variable quantity of total zinc has been added. As noted in chapter one, a Bq of short-lived radionuclide such as [65]Zn represents only a very small quantity of the element, in fact 1 Bq l^{-1} (27 pCi l^{-1}) [65]Zn is only $3.2 \times 10^{-9} \mu g \, l^{-1}$ zinc; the amount of stable zinc in coastal sea water is usually about 2 to 5 $\mu g \, l^{-1}$. The fronds of *Laminaria* also appear to adsorb relatively more zinc from low than from high sea water concentrations, and thus alter the specific activity of previously accumulated [65]Zn. Higher zinc concentrations are also attained by slower-growing weed.

Different algal species, even in the same area, can be shown to accumulate the same radionuclide to quite different levels. The intertidal species listed in table 6.3 were all collected at the same time, fairly close to Windscale. The data have been normalized to those of *Porphyra* in order to illustrate more readily the inter-species differences. The extent to which each radionuclide is accumulated over the level of that in the sea water reflects, in part, the extent to which each radionuclide is conservative to sea water; the relationship is an inverse one. Thus ^{137}Cs is accumulated by *Porphyra* (on a wet weight basis) to a level some 10 times, [95]Zr/[95]Nb 400 times, [144]Ce 1000 times, and

TABLE 6.3 *Radionuclide concentrations in different species of marine algae, normalized to those of the edible species Porphya, collected in the Windscale area of the Irish Sea.*

Species	^{106}Ru	^{95}Zr /^{95}Nb	^{144}Ce	^{137}Cs	^{239}Pu + ^{240}Pu	^{241}Am
Rhodophyceae						
Porphyra						
umbilicalis	1·0	1·0	1·0	1·0	1·0	1·0
Rhodymenia						
palmata	0·9	2·6	2·7	7·4	3·4	3·5
Chondrus crispus	0·8	1·8	3·2	2·4	4·0	4·3
Chlorophyceae						
Enteromorpha						
spp.	1·0	5·0	4·6	6·2	7·3	6·6
Cladophora spp.	2·5	13·4	11·3	4·5	16·9	18·7
Ulva lactuca	1·8	10·3	6·9	5·1	21·6	18·7
Chaetomorpha						
spp.	1·2	8·0	6·1	8·9	—	—
Phaeophyceae						
Fucus spiralis	0·1	1·3	0·9	3·4	4·7	2·4
Fucus vesiculosus	0·2	1·6	0·8	3·9	5·0	2·2
Ascophyllum						
nodosum	0·1	1·0	0·5	2·8	3·6	1·2
Fucus serratus	0·4		1·2	3·6	8·2	3·1
Laminaria digitata	0·2	1·2	0·6	3·2	3·9	1·7

From Pentreath, R.J., 1976, *Monitoring of Radionuclides.* In *FAO Fisheries Technical Paper* No. 150, FIRI/T150, p.8; Crown copyright, reproduced by permission.

^{106}Ru 1500 times as great as the concentrations prevailing in the sea water.

Three species of *Fucus* are listed in table 6.3 and this genus is used regularly for the monitoring of radionuclides around the shores of the British Isles. In addition to the nuclides listed in the table, fucoids have been used for the monitoring of ^{90}Sr, ^{110m}Ag, ^{65}Zn and ^{60}Co. In fact, the routine analysis of fucoids revealed the presence of another radionuclide, ^{99}Tc, in the Windscale area which, because of the extremely low concentrations, would not otherwise have been seen at all. It had not been detected in the routine analyses of *Porphyra*.

Adsorption is also the principal means of radionuclide accumulation by phytoplankton, although some radionuclides, particularly those of essential trace elements, are absorbed. Concentration factors over the sea water exceed 10^3 for a number of elements, including those with radionuclides formed as fission products – zirconium, niobium, cerium, ruthenium – and those with radionuclides formed as neutron activation products, including zinc, manganese, cobalt, iron, chromium, nickel, silver and plutonium.

Phytoplankton is eaten by zooplankton, a term which covers examples from a wide range of phyla. Depending on the time of year, zooplankton may contain the larvae of benthic animals – polychaetes, molluscs, decapods, echinoderms – in addition to representatives of the major phyla, from protozoa to the vertebrates. By and large zooplankton display lower concentration factors for the majority of radionuclides than do phytoplankton, even though they are exposed to them both through their food – which includes not only phytoplankton but other zooplankton – and by adsorption from sea water. This is partly the result of relatively poor assimilation efficiencies across the gut wall, the assimilation efficiency for carbon being about 50%. Fission products are poorly retained from food by euphausiids and copepods, elements such as cerium virtually passing straight through the animal. A number of neutron activation products have been studied and some of these are absorbed from food, notably ^{65}Zn and ^{54}Mn. Planktonic crustacea may lose the radionuclides which they have accumulated either via their excretory products or through the process of moulting. As these crustacea have a fairly large surface-

TABLE 6.4. *Loss of some trace elements to the sea water, by way of moults and faecal pellets, by the euphausiid Meganyctiphanes norvegica*

Element	% body burden in moult (calculated on a dry weight basis)	Daily loss (μg kg^{-1}) by euphausiid per day via:	
		Faecal pellets	Moults
Ag	31	80	26
Co	34	130	7
Cr	48	1400	48
Fe	28	910000	2100
Mn	21	9200	110
Zn	18	36000	1300
Ce	44	7600	11
Cs	2	230	0·2
Sr	23	3000	3200

From Fowler, S.W., 1977, *Nature, London*, 269, 51; courtesy of the author and Macmillan Journals Ltd.

to-volume ratio, their exoskeletons provide an excellent surface for adsorption. Radionuclides which adsorb in this way, such as a number of the fission products, are not the only ones excreted via moulting, however, because some of those assimilated internally may also be deposited into the cast exoskeleton.

One particular euphausiid, *Meganyctiphanes norvegica*, has been studied in detail. The data given in table 6.4 are a selection from a more detailed examination of trace elements in this species. The study was made in the Mediterranean, where *M. norvegica* moults every few days. The moult consists only of the thin outer layer of the exoskeleton, which represents an average 7·7% of the organism's dry weight. Faecal production is higher than the production of cast moults, and the former route is considered to be particularly important in transferring elements from surface layers to the sea bed. It has been observed that faecal pellets decompose more slowly when sinking through the water column than do the moults. Zooplanktonic animals such as euphausiids are very seasonal in abundance in the coastal waters of temperate regions, however, and their relative effect on the transportation of radionuclides, in comparison to direct physical processes, has yet to be evaluated.

The pelagic food chain passes from zooplankton to fish, and it is here that the greatest difference between those radionuclides which can adsorb on to surfaces and those which can be absorbed across the gut wall, or gill, is most pronounced. Analyses made in the Pacific early in the nuclear weapons test programmes very quickly illustrated such differences. Plankton samples contained a mixture of the fission product radionuclides, such as ^{106}Ru and ^{95}Zr/^{95}Nb, and the neutron activation products ^{57}Co, ^{58}Co, ^{60}Co, ^{55}Fe, ^{59}Fe, ^{65}Zn and ^{54}Mn. Liver samples of flying fish taken in the same area contained principally ^{55}Fe and ^{59}Fe, whereas muscle samples contained some ^{55}Fe and ^{59}Fe but predominatly ^{65}Zn. Both liver and muscle also contained the cobalt isotopes but the other radionuclides were not detected at all. A further difference was observed in tuna, which prey upon flying fish. Although ^{55}Fe and ^{59}Fe were present in tuna livers it was the ^{65}Zn which predominated, as it also did in muscle samples; only slight traces of the cobalt isotopes could be detected.

A similar reduction in the number of radionuclides detected has been observed in fish caught within the vicinity of nuclear power stations and reprocessing plants. In fact virtually the only gamma-emitting radionuclides detected in fish caught around the United Kingdom are those of an element which has not been discussed above – caesium. Caesium is accumulated by phytoplankton and by zooplankton, the latter retaining it from both food and water and storing it in the soft tissues rather than in the exoskeletons. Such accumulation by small planktonic crustacea nevertheless results in concentrations which are only about one order of magnitude greater than that of the sea water. Surprisingly the concentration factors of ^{134}Cs and ^{137}Cs in fish are higher, not less, than those of its food.

Before looking at fish in more detail, it is as well to remember that in coastal waters many species of fish do not obtain their food from plankton but from the benthos. And in view of so many radionuclides eventually being deposited on to the sea floor, the benthic fauna is also an important link in understanding the cycling of radionuclides in the sea. The benthic fauna is also less seasonal in abundance than the planktonic populations. As noted above, plankton in the coastal waters of temperate regions show very marked cycles of productivity, being present in immense numbers in the early summer, and again in the autumn, and yet virtually absent for the rest of the year. The benthos contains a high proportion of annelids, molluscs and echinoderms, and of these the molluscs have been the subject of numerous studies on radionuclide accumulation. Bivalve molluscs are largely filter-feeding and display a remarkable ability to take up a wide range of radionuclides, both fission and neutron activation products. These are usually stored in the hepatopancreas but other organs, such as the kidney or other excretory organ, may specifically accumulate certain radionuclides. The ability of different species to accumulate specific radionuclides has made them particularly useful as indicators. Scallops (*Pecten maximus*), for reasons unknown, attain very high concentration factors relative to sea water for manganese, which is highly concentrated in the kidneys. Oysters, on the other hand, as discussed briefly in chapter 5, have very high concentration factors for zinc. Mussels have concentration factors for iron which are considerably higher than either manganese or zinc. The iron is particularly highly concentrated in the hepatopancreas but appears to be excreted through the production of byssus threads – which attach the mussel to the substrate – via a gland in the foot.

Benthic crustacea, such as crabs and lobsters, also display a remarkable ability to concentrate certain elements; and they also display a remarkable ability to regulate the body burdens of some of them. The crab *Carcinus maenas* can maintain a comparatively constant zinc content. Zinc can be accumulated directly from sea water – via the gills – by a process which involves the binding of zinc to the blood. The turnover of zinc takes place more quickly in sea water which has a high zinc content and thus the excretion of ^{65}Zn can be increased by placing crabs in high-zinc content sea water. Zinc can also be absorbed very efficiently from the gut of crabs and lobsters and this, in fact, is probably the major route of uptake.

Both molluscs and crustacea absorb radionuclides from food to a higher degree than the gut of vertebrates; their gut morphologies are also markedly different. An excellent example of such a difference in retention is that of the absorption of plutonium fed to crabs and fish. The radionuclide used in each experiment, ^{237}Pu, is not one that occurs in the environment; but it is a gamma-emitter – which permits whole-body counting of the animal – and it can be used in the laboratory at similar gravimetric concentrations to those of ^{239}Pu + ^{240}Pu in contaminated environments. The difference in retention is shown in figure 6.5. Polychaete worms (*Nereis diversicolor*)

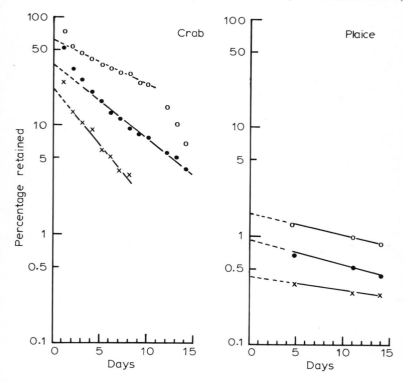

Figure 6.5. Percentage ^{237}Pu(IV) retained by crabs (*Carcinus maenas*), and plaice (*Pleuronectes platessa*), fed labelled *Nereis diversicolor*. Crab data from Fowler, S. W., and Guary, J. C., 1977, *Nature, London*, **266**, 827, courtesy of the authors and Macmillan Journals Ltd; plaice data from Pentreath, R. J.; Crown copyright, reproduced by permission.

which had accumulated ^{237}Pu(IV) from sea water were fed, in this experiment, to three *Carcinus maenas*. The crabs retained from 20 to 60% of the ^{237}Pu initially given, and over half of this was found to accumulate in the digestive gland. In contrast, plaice (*Pleuronectes platessa*) fed ^{237}Pu(IV)-labelled *Nereis* retained very little (<1·5%) of the radionuclide initially given, and none could be detected in any internal organ other than that adhering to the gut wall.

The plaice does appear to be able to accumulate plutonium direct from sea water to a very limited extent, however, but for the majority of the radionuclides which are accumulated by fish the major pathway appears to be the food chain. This is certainly true of the majority of neutron activation products which have been studied. Only a few of the longer lived fission products are accumulated at all by fish, even though their concentrations in food species may be very high. This is especially the case for a bottom feeding fish such as the plaice, which not only eats benthic organisms – which accumulate a wide range of radionuclides – but also ingests the contaminated sediment adhering to them. Analyses of muscle samples of Irish·Sea plaice by gamma spectrometry reveals only the presence of caesium, whereas the gut contents contain a number of other radionuclides (table 6.5); ^{106}Ru, however, can be detected in the liver. Caesium can also be accumulated directly from sea water and for the plaice this pathway accounts for approximately half the fish's body burden. A concentration factor of about 40 is attained in adult plaice.

There have been many studies on the accumulation of radionuclides by marine fish and a number of them illustrate the difficulties of interpreting, and predicting, the behaviour of radionuclides in aquatic environments. With the exception of the few marine mammals, all animals living in the sea are poikilothermic – they cannot regulate their body temperatures, or at least only to a limited degree. In temperate regions which experience quite a wide range of temperature, this results in a seasonal variation in food intake, and thus in growth. Fish do not grow to maturity and then stop, but continue to grow throughout life – although the rate of growth does decrease at some point. The rate of accumulation of radionuclides is dependent upon body size, as can be demonstrated for the uptake of ^{137}Cs by the plaice (figure 6.6). Fish of different sizes also have different rates of excretion, and thus different biological half-times. Whereas the rate of intake decreases with increasing body size, the biological half-times tend to increase with an increase in body size (figure 6.7) and the net result is, to some extent, for the two effects to balance each other out. Just as important as body size to radiocaesium accumulation is the

TABLE 6.5. Radionuclide concentrations (wet weight) in plaice (Pleuronectes platessa) of different ages taken in the vicinity of Windscale, Irish Sea, in 1968/1969.

Age group	Flesh	Gut contents			
	^{137}Cs mBq g^{-1} (pCi g^{-1})	^{144}Ce mBq g^{-1} (pCi g^{-1})	^{106}Ru mBq g^{-1} (pCi g^{-1})	^{137}Cs mBq g^{-1} (pCi g^{-1})	^{95}Zr/^{95}Nb mBq g^{-1} (pCi g^{-1})
0	59·2 (1·60)				
I	69·9 (1·89)	1150·7 (31·1)	5150·4 (139·2)	181·3 (4·9)	5794·2 (156·6)
II	25·9 (0·70)	880·6 (23·8)	1343·1 (36·3)	44·4 (1·2)	4199·5 (113·5)
III	32·9 (0·89)	932·4 (25·2)	1509·6 (40·8)	70·3 (1·9)	3193·1 (86.3)
IV	40·0 (1·08)	902·8 (24·4)	2090·5 (56·5)	62·9 (1·7)	3122·8 (84·4)

From Pentreath, R. J., Jefferies, D. F., and Woodhead, D. S., 1973, Radionuclides in Ecosystems, Proceedings of a Symposium, CONF-710501-P2, USAEC, p.731; Crown copyright, reproduced by permission.

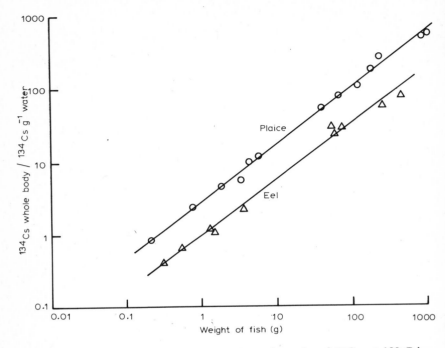

Figure 6.6.   Relationship between weight and uptake of ¹³⁴Cs, at 10° C by plaice (*Pleuronectes platessa*) and eel (*Anguilla anguilla*). From Morgan, F., 1964, *Journal of the Marine Biological Association of the United Kingdom*, **44**, 259; Crown copyright, reproduced by permission.

rate of change of body size, i.e. growth rate. The temperature of the sea water also has a direct effect on the rate of intake and excretion of radionuclides by fish of the same size because it affects their overall rate of metabolism, including the rate at which water is passed over the gills.

All of these effects relate to the accumulation of a radionuclide direct from the sea water. To these must be added the effects of diet, efficiency of retention from different dietary components, different conversion efficiencies of food to fish flesh, and so on. As if all of these parameters were not enough to contend with, it needs to be remembered that the majority of radioactive discharges are made from a single point of release and thus their concentrations in the environment, as previously illustrated in figures 6.1 and 6.2, decrease with increasing distance from the source. Fish – and other motile organisms – therefore experience fluctuating concentrations within a

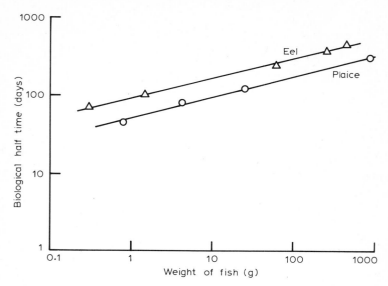

Figure 6.7. Relationship between biological half-time of [134]Cs and weight, at 10° C, for the plaice (*Pleuronectes platessa*) and the eel (*Anguilla anguilla*). From Morgan, F., 1964, *Journal of the Marine Biological Association of the United Kingdom*, **44**, 259; Crown copyright, reproduced by permission.

relatively small area even if the rate of input were constant, which it usually is not. It is, nevertheless, usually possibly to develop some form of model to describe and predict concentrations in marine organisms, even fish. As discussed in chapter 5, for the purposes of regulating discharges it is not always necessary to understand every detail of how a radionuclide is accumulated; it is sufficient to know what sort of concentrations will result from a unit rate of discharge. The example given in chapter 5 was for an attached algae, but to illustrate that even for fish the concentrations attained can be related remarkably well to rates of discharge, table 6.6 gives data for [238]Pu + [239]Pu + [240]Pu and [241]Am in plaice caught near Windscale over a 2 year period. Despite all of the parameters which are known to affect radionuclide accumulation, the data in the table are remarkably constant.

The sea is chemically relatively constant in comparison with estuarine and freshwater environments. It is relatively stable with regard to mineral composition, temperature, density and pH. Conditions in freshwater are much less constant, particularly with regard to elemental concentration and ionic composition. Freshwaters are

TABLE 6.6. *Relationship between concentrations of ^{238}Pu plus $^{239}Pu + ^{240}Pu$, and ^{241}Am in the muscle of plaice (Pleuronectes platessa) caught in the vicinity of Windscale, Irish Sea, and the daily quantities discharged. Results may be read either as $fBq\ g^{-1}$ wet per $Bq\ d^{-1}$ discharged, or as $fCi\ g^{-1}$ wet per $Ci\ d^{-1}$ discharged.*

Month of collection	$^{238}Pu + ^{239}Pu + ^{240}Pu$	^{241}Am
1975		
February	0·06	0·24
May	0·09	0·24
August	0·18	0·33
November	0·12	0·27
1976		
February	0·09	0·60
May	0·36	1·35
August	0·09	0·21
November	0·03	0·63
1977		
February	0·03	0·63

From Pentreath, R. J., and Lovett, M. B., 1978, *Marine Biology*, 48, 19; Crown Copyright, reproduced by permission.

therefore more exacting environments in which to live and consequently they contain a much smaller variety of species than the sea. In many ways estuaries are even more difficult environments in which to live because their physical and chemical parameters fluctuate to greater extremes, and do so daily with the tidal ebb and flow. Estuaries also receive a large input of elements from the land as freshwater run-off, and are efficient sediment traps. In fact, the physical and chemical processes which occur in estuaries are very complex indeed. In order to demonstrate the differences which can occur in moving from the marine, to estuarine, to freshwater environments, it will be convenient to continue with the observations which have been made on caesium. The blood of marine animals is more or less on an osmotic par with sea water, but freshwater is considerably lacking in the monovalent elements which predominate in sea water. Consequently there is a tendency for water to flow into freshwater organisms, and for salts to flow out down a concentration gradient. Both marine and freshwater organisms are adapted physiologically to these two sets of conditions, whereas estuarine animals and plants have to contend with fluctuating salinities. They may accomplish this either by displaying a remarkable ability to adjust, chemically, to the fluctuating conditions or, equally remarkably, display an amazing tolerance to the fluctuating chemical conditions. Marine invertebrates in general absorb caesium more slowly than potassium, but the former attains a slightly higher concentration factor. Concentration factors for both elements are usually in the range 1 to 20. In brackish water the concentration of potassium is decreased relative to sea water, but many animals manage to maintain blood and tissue potassium concentrations which are only slightly reduced, and thus very much higher than those of the medium. Caesium is chemically very similar to potassium and therefore under these conditions it is also concentrated to a much higher level relative to the water. A good example is that of the brackish water isopod *Sphaeroma hookeri* (figure 6.8) which can survive dilutions down to about 2·5% of full sea water strength. Equilibrium concentration factors for ^{137}Cs are about 7 in 100% sea water but rise to 200 to 300 in 2·5% sea water.

The ^{137}Cs concentration factors of *Sphaeroma hookeri* are attained by direct accumulation from sea water; ^{137}Cs is also well absorbed from food. It was noted earlier that the marine fish, the plaice, attains half of its body burden of ^{137}Cs by direct uptake from sea water. This direct uptake results in a concentration factor of about 20. The environmental concentration factor for caesium in the freshwater brown trout (*Salmo trutta*) is of the order of a thousand. Experiments on the direct accumulation of ^{137}Cs from water by trout, however, have shown that the concentration factors attained are even lower than those of plaice. It is therefore apparent that the much higher ^{137}Cs content of the trout flesh relative to that of freshwater is the result of ^{137}Cs retained from food.

There is, nevertheless, a relationship between the potassium content of the water and the ^{137}Cs content of the fish. Trout taken from a number of freshwater areas in the British Isles during 1965 and 1966 were found to contain ^{137}Cs resulting from atmospheric fallout, the concentrations correlating inversely with the potassium content of the water (figure 6.9). A similar story can be told for ^{90}Sr: the same survey analysed the fish flesh for this radionuclide, and its concentration was found to be inversely correlated with the calcium content of the water.

All of these studies are but the briefest examples of the complexities relating to the transfer of radionuclides through aquatic ecosystems. In some instances the observed data can be inferred, or interpreted, from available physiological knowledge. In a number of cases, however, our basic knowledge is insufficient. This is obviously a handicap in predicting the longer term behaviour of many radio-

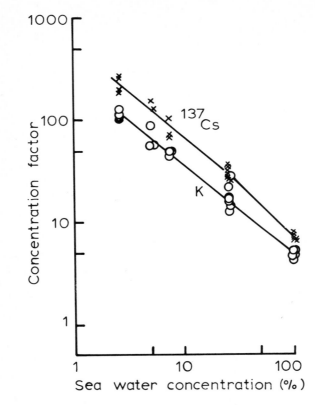

Figure 6.8. Comparison of equilibrium concentration factors for ^{137}Cs with concentration factors for potassium which are maintained by the isopod *Sphaeroma hookeri* in diluted sea water at 20° C. From Bryan, G. W., 1963. In *Nuclear Detonations and Marine Radioactivity*, edited by S. H. Small (Norwegian Defense Research Establishment), Kjeller: 153; courtesy of Dr G. W. Bryan.

nuclides and there is much research to be done. But it should also be appreciated that much research has been done, and that this has provided a wealth of data on the behaviour of many trace elements in the environment, data which otherwise would not have been forthcoming.

6.4. *Radiation effects on the environment*

Little has been said in the preceding sections of the actual concentrations observed in the environment as a result of the authorized discharges of radioactive wastes. By and large they lie in the mBq

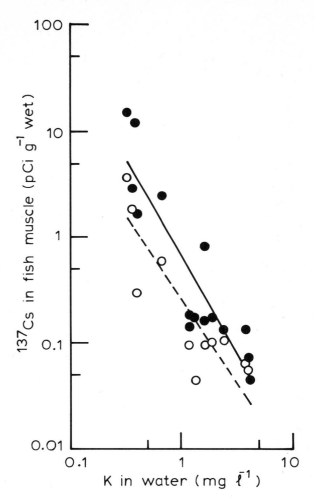

Figure 6.9. Relationship between ^{137}Cs in brown trout (*Salmo trutta*) and the potassium concentration in the water of a number of United Kingdom rivers in 1965 (●) and 1966 (○). From Preston, A., Jefferies, D.F., and Dutton, J. W. R., 1967, *Water Research*, **1**, 475; Crown copyright, reproduced by permission.

(pCi) range, such concentrations resulting solely from calculations based on minimizing the exposure to man. Knowledge of such concentrations in environmental samples, if not for calculating exposure to man, is only useful if it is to be used to estimate the radiation doses received by the plants and animals themselves, and by the relating of

TABLE 6.7. *Summary of maximum estimated background dose rates to some marine organisms.*

Source	Phytoplankton† μGy h⁻¹ (μrad h⁻¹)	Zooplankton† μGy h⁻¹ (μrad h⁻¹)	Mollusca‡ μGy h⁻¹ (μrad h⁻¹)	Crustacea‡ μGy h⁻¹ (μrad h⁻¹)	Fish‡ μGy h⁻¹ (μrad h⁻¹)
NATURAL BACKGROUND					
Cosmic radiation	0·005 (0·5)	0·005 (0·5)	0·005 (0·5)	0·005 (0·5)	0·005 (0·5)
Internal radioactivity	0·073 (7·3)	0·157 (15·7)	0·149 (14·9)	0·214 (21·4)	0·042 (4·2)
Water radioactivity	0·004 (0·4)	0·002 (0·2)	0·001 (0·1)	0·001 (0·1)	0·001 (0·1)
Sediment radioactivity γ	—	—	0·160 (16·0)	0·160 (16·0)	0·160 (16·0)
β	—	—	0·210 (21·0)	0·210 (21·0)	0·210 (21·0)
FALLOUT					
Internal radioactivity (³H, ⁹⁰Sr, ¹³⁷Cs, ²³⁹Pu)	0·009 (0·9)	0·134 (13·4)	0·003 (0·3)	0·001 (0·1)	0·001 (0·1)
Other internal nuclides	0·246 (24·6)	1·340 (134·0)	0·007 (7·7)	0·004 (0·4)	0·017 (1·7)

† At 20 m depth, remote from sea bed. ‡ At 20 m depth, on the sea bed.

From IAEA Technical Report Series No. 172, 1976; courtesy of the International Atomic Energy Agency, Vienna.

TABLE 6.8. *Summary of maximum estimated background dose rates to some freshwater organisms.*

Source	Phytoplankton† (μGy h⁻¹)	(μrad h⁻¹)	Zooplankton† (μGy h⁻¹)	(μrad h⁻¹)	Mollusca‡ (μGy h⁻¹)	(μrad h⁻¹)	Crustacea‡ (μGy h⁻¹)	(μrad h⁻¹)	Fish‡ (μGy h⁻¹)	(μrad h⁻¹)
NATURAL BACKGROUND										
Cosmic radiation	0·027	(2·7)	0·027	(2·7)	0·022	(2·2)	0·022	(2·2)	0·022	(2·2)
Internal radioactivity	?	?	?	?	?	?	?	?	0·048	(4·8)
Water radioactivity	0·062	(6·2)	0·009	(0·9)	0·004	(0·4)	0·004	(0·4)	0·004	(0·4)
Sediment radioactivity γ	—		—		0·160	(16·0)	0·160	(16·0)	0·160	(16·0)
β	—		—		0·210	(21·0)	0·210	(21·0)	0·210	(21·0)
FALLOUT										
Internal radioactivity	?	?	?	?	0·002	(0·2)	?	?	0·264	(26·4)
Water radioactivity	0·005	(0·5)	0·004	(0·4)	0·001	(0·1)	0·001	(0·1)	0·001	(0·1)
Sediment radioactivity γ	—		—		0·058	(5·8)	0·058	(5·8)	0·058	(5·8)
β	—		—		0·035	(3·5)	0·035	(3·5)	0·035	(3·5)

†At 1 m depth >1 m from river bed. ‡At 2 m depth, on the river bed.
? = no data available.

From IAEA Technical Report Series No. 172, 1976; courtesy of the International Atomic Energy Agency, Vienna.

these values to absorbed doses which are known to cause them harm. But one must again remember that all living organisms are exposed to many sources of background radiation. It is, therefore, particularly useful first of all to examine the estimations which have been made with regard to such background dose rates, and then to compare these with areas which receive substantial quantities of discharged radioactivity. Because the aquatic environments receive the bulk of the longer lived radionuclides it is as well to begin here. Marine organisms receive background radiation from both natural sources and from fallout. A summary of absorbed dose rates for idealized examples of phytoplankton, zooplankton, molluscs, crustacea and fish are given in table 6.7. The calculations were made using the assumptions that phytoplankton were spherical and 50 μm diameter, that zooplankton were essentially cylinders 0·5 cm long and 0·2 cm diameter and that molluscs, crustacea and fish were cylinders of 4, 6 and 10 cm diameter and 1, 15 and 50 cm long, respectively. The table was compiled, using all of the data available at the time, by a panel convened by the IAEA in 1971 and 1974.† In the marine environment the incorporation of radionuclides into organisms, particularly the natural alpha emitter ^{210}Po, contributes the largest fraction of the dose rate. Cosmic radiation is especially important for animals actually living at the surface of the sea, where the dose rate is about 0·4 μGy h^{-1} (4 μrad h^{-1}); it decreases to 0·005 μGy h^{-1} (0·5 μrad h^{-1}) at 20 m and becomes negligible at 100 m depth. The ^{40}K in sea water, at about 300 pCi l^{-1}, contributes a substantial proportion of the external dose rate to pelagic organisms but the natural radionuclides in sediments – which vary in the same way as those in soils, as shown in chapter 2 – contribute the major fraction of the external dose rate to animals on the sea bed.

Similar data (table 6.8) have been compiled for freshwater organisms. Freshwater contains appreciable quantities of ^{222}Rn and its daughters and these contribute to the external dose rate, particularly for phytoplankton. A major source of internal dose rate for fish is their accumulated ^{40}K, and although data are lacking for the other freshwater organisms, they are likely to be of the same order. By and large the total natural radiation dose rates in the two environments are very similar. Fallout radionuclides have also contributed to the background radiation in the aquatic environment and these data are also presented in the tables.

The next step is to repeat the dose rate calculations for those areas

† IAEA Technical Report Series, No. 172.

TABLE 6.9. *Summary of maximum estimated dose rates to some aquatic organisms resulting from waste disposal practices.*

Source	Phytoplankton†		Zooplankton†		Mollusca‡		Crustacea‡		Fish‡	
	µGy h⁻¹	(µrad h⁻¹)	µGy h⁻¹	(µrad h⁻¹)	µGy h⁻¹	(µrad h⁻¹)	µGy h⁻¹	(µrad h⁻¹)	µGy h⁻¹	(µrad h⁻¹)
MARINE DISPOSAL **WINDSCALE (1968)**										
Internal radioactivity	21·00	(2100)	69·00	(6900)	0·59	(59)	0·68	(68)	0·02	(2)
Water radioactivity	0·03	(3)	0·03	(3)	0·01	(1)	0·01	(1)	0·01	(1)
Sediment radioactivity γ	—	—	—	—	33·40	(3340)	33·40	(3340)	33·40	(3340)
β	—	—	—	—	53·80	(5380)	53·80	(5380)	53·80	(5380)
FRESHWATER DISPOSAL **COLUMBIA RIVER (1964-1966)**										
Internal radioactivity	1·95	(195)	150·00	(15000)	240·00	(24000)	?	?	21·00	(2100)
Water radioactivity	0·03	(3)	0·03	(3)	0·01	(1)	0·01	(1)	0·01	(1)
Sediment radioactivity γ	—	—	—	—	8·60	(860)	8·60	(860)	8·60	(860)
β	—	—	—	—	0·29	(29)	0·29	(29)	0·29	(29)

† at 20 m depth, remote from sea bed, for marine disposal; and at 1 m depth but >1 m from river bed for freshwater disposal.
‡ at 20 m depth, on the sea bed, for marine disposal; and at 2 m depth, on the river bed, for freshwater disposal.
? = no data available.

From IAEA Technical Report Series No. 172, 1976; courtesy of the International Atomic Energy Agency, Vienna.

receiving relatively large inputs of radioactive wastes by using known environmental concentrations, and concentration factors, for the radionuclides discharged. For the marine environment the obvious choice is the Windscale area (table 6.9). For the benthic organisms the overriding source of irradiation is that from the radionuclides adsorbed to the sediments. The only freshwater site comparable to that of Windscale is the Columbia river, and that only in the past tense because the large discharges of radioactivity into the river have now ceased, seven of the reactors at the Hanford plutonium production plant having been closed down. The data from the Columbia river, in the USA, are the best available, however, and these have been used to calculate the values given in table 6.9. As can be seen, the major source of irradiation was from radioactivity absorbed by the organisms; this was principally ^{32}P.

All of these data are theoretical calculations, but there are data which have been derived by direct *in situ* measurements. An example is that of a study of the absorbed dose rate by plaice in the Irish Sea. Over 3500 plaice were caught and thermoluminescent dosimeters (TLDs) were attached to both the underside and topside of the fish by means of a standard Petersen disc tag, and the fish returned to the sea. Approximately a third of these fish were subsequently caught as a result of commercial fishing operations, and the TLDs returned for analyses. The dose rates recorded for the underside of the fish gave a mean value of $3.5\,\mu$Gy h^{-1} (0.35 mrad h^{-1}) but displayed a considerable, log-normal distribution range (figure 6.10). The mean ratio of the response of the upper dosimeter to the lower dosimeter was 0.68, but the range was from 0.13 to 1.79, the higher values presumably reflecting the characteristic behaviour of this species; it buries itself into the top layers of the sediment.

A difficulty in studying larger organisms such as fish is deciding which organ or tissue to select for the purposes of calculating an absorbed dose. Because it is known that the gonad is particularly radiosensitive, this organ was selected for the plaice studies; its absorbed dose was considered to be derived from the radionuclides on the sea bed, from the radionuclides in the sea water, from those few radionuclides which are actually assimilated into the fish – principally ^{137}Cs – and from those radionuclides which, although not assimilated, are ingested and pass along the gastrointestinal tract from where they can irradiate the gonad. The estimates of the dose rate to the lower gonad of plaice living in their first full calendar year (I-group) in the vicinity of Windscale are given in table 6.10.

The embryonic stage is also relatively radiosensitive and estimates

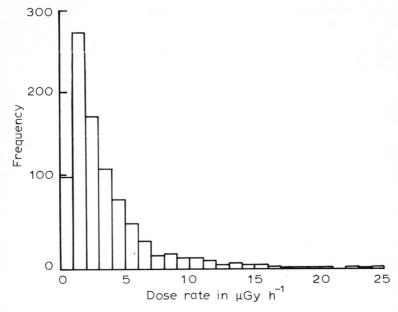

Figure 6.10 Dose rate to plaice (*Pleuronectes platessa*) in the north-east Irish Sea as indicated by the dosimeter on the underside of the fish. From Woodhead, D. S., 1973, *Health Physics*, **25**, 115; Crown copyright, reproduced by permission.

TABLE 6.10. *Estimates of the dose rate to the lower gonad of a I-group plaice (Pleuronectes platessa) in the Windscale area, Irish Sea, in 1969.*

Source	Dose rate			
	β radiation		γ radiation	
	μGy h^{-1}	(μrad h^{-1})	μGy h^{-1}	(μrad h^{-1})
Water	0·0005	(0·05)	0·0023	(0·23)
Seabed silt	7·3	(730·0)	7·0	(700·0)
Tissue	0·012	(1·2)	0·0021	(0·21)
Gut contents	0·3	(30·0)	0·0065	(0·65)

From Pentreath, R. J., Jefferies, D. F., and Woodhead D. S., 1973, *Radionuclides in Ecosystems*. Proceedings of a Symposium, CONF-710501-P2, USAEC, p.731; Crown copyright, reproduced by permission.

of the dose rate received by plaice eggs – which float in the sea water for a period of just over 2 weeks before the larvae emerge – have also been calculated for the major gamma- and beta-emitting radio-

TABLE 6.11.   *Ranges of acute lethal radiation doses for adults of a number of aquatic organisms.*

Organism	Gy		(krad)		Type of experiment†
Bacteria	45 —	7350	(4·5 —	735)	LD_{90}
Blue-green algae	<4000 —	>12000	(<400 —	>1200)	LD_{90}
Other algae	30 —	1200	(3 —	120)	LD_{50}
Protozoa	? —	6000	(? —	600)	LD_{50}
Mollusca	200 —	1090	(20 —	109)	$LD_{50/30}$
Crustacea	15 —	566	(1·5 —	56·6)	$LD_{50/30}$
Fish	11 —	56	(1·1 —	5·6)	$LD_{50/30}$

†LD_{90} = lethal dose for 90% of the population; LD_{50} = lethal dose for 50% of the population; and $LD_{50/30}$ = lethal dose for 50% of the population after 30 days.

From IAEA Technical Report Series No. 172, 1976; courtesy of the International Atomic Energy Agency, Vienna.

nuclides present in the Irish Sea. Their combined dose rate was estimated as being less than that received from [40]K.

What is the likely effect of such absorbed doses? One can start by asking how large a dose is required to kill the animal. Doses required to kill 50% of the population are termed LD_{50} data – LD, for lethal dose. An indication of the range of values obtained as a result of such experiments are given in table 6.11 for aquatic organisms. The data for molluscs, crustacea and fish are for $LD_{50/30}$ experiments, i.e. from those experiments which have derived the doses at which 50% of the irradiated population have died within 30 days. This raises another important question – survival after how long a period? The 30-day limit has usually been chosen for its convenience, and for its comparability with data on small mammals such as mice which, if surviving a 30-day period, have a good chance of surviving for their normal life span. Nevertheless, the doses required are enormous and of little relevance to environmental levels. Such data could be considerably improved by deriving LD_{50} values over longer periods of time; periods which relate to the longevity of the species. It has been observed that the LD_{50} for the estuarine fish *Fundulus heteroclitus*, at 5°/₀₀ and 27° C, is 24·5 Gy (2450 rad) after 20 days, but drops markedly to 3·5 Gy (350 rad) after 60 days. Other studies have confirmed these trends. Experiments with the rough-skinned newt (*Taricha granulosa*) derived a $LD_{50/30}$ dose of 30 kR, but when the experiment was continued for a period of 300 days it was found that the *minimal* LD_{50} – the LD_{50} for a dose range at which survivors can be expected to live for a prolonged period of time – was not greater than 250 R. It is evident that for experiments with poikilothermic animals, sufficient time must be allowed for the induced radiation damage fully to take effect.

When organisms are irradiated continuously at low dose rates a greater total dose is required to produce the same degree of damage. This is because at low dose rates there is a competition, at tissue level, between the injury and the mechanisms of repair. Such chronic exposure is more relevant to the contaminated environment, and a large number of such experiments have been made. Using external sources, for example, total gamma doses of 0·6 to 500 R, at rates of 10 mR h $^{-1}$ up to 1 R h $^{-1}$, were delivered to plaice eggs from fertilization to hatching: no significant differences in survival or abnormalities were observed. Such dose-rates are well in excess of those encountered in the Windscale environment. A number of experiments on the effects of radionuclides in the water – particularly in connection with fish eggs – and on radionuclides incorporated into body tissues have also been made. Studies with tritium, which is incorporated into the tissues, have demonstrated that concentrations in the region of 18·5 GBq l^{-1} (0·5 Ci l^{-1}) are required before significant effects are observed on developing fish embryos. Adult fish (*Salmo gairdnerii*) have been fed daily with high concentrations of different radioisotopes to observe any deleterious effects. Daily intakes of 2·22 kBq (0·06 μCi) ^{32}P g $^{-1}$ wet of fish for 4 months resulted in leucopaenia, but daily intakes of 18·5 kBq (0·5 μCi) ^{90}Sr/ ^{90}Y g $^{-1}$ wet and 370 kBq (10 μCi) ^{65}Zn g $^{-1}$ wet were required to achieve the same effect.

All of the data discussed above relate to effects on individuals, and yet it is the effect on the population which should be of the greatest concern. As with human populations it is possible to derive a statistical probability that, as a result of the enhanced levels of radioactivity in the aquatic environment, one animal, or one fish egg, in *n* thousand may die. For highly fecund species, such as many of the teleosts, over 99·9% of eggs and larvae will not survive to maturity in any case because of natural predation. Such considerations may be considered as pure theory, however, and it is thus of special interest to review briefly experiments which have been made on chinook salmon (*Oncorhynchus tshawytscha*). This species, in common with other salmon, is very fecund and also migratory. Long-term studies have been made by delivering chronic low-level gamma irradiation to the eggs during embryonic development. Initially dose-rates of 0·5 to 20 R d⁻¹ were delivered from shortly after fertilization until the commencement of feeding. The small fish, called fingerlings, were then reared and allowed to migrate to the sea. During the second year, precocious males returned to the hatchery, and in the third and fourth years both male and female adult fish returned. Various crosses of the returned fish were made and some of the resultant eggs and larvae were also irradiated. Large numbers of fish have been used in these experiments – 96 000 to 250 000 fingerlings being released per experiment – and the results indicate that at the dose rates used, although abnormalities in young fish were increased at all dose rates, the number of adult fish returning to breed was not affected. In fact, no damage to the fish stock, sufficient to reduce the reproductive capability over a period of several generations, was produced. It was even observed that the low-dose irradiated stock returned in greater numbers, and produced a greater total of viable eggs, than the control stock fish. At dose rates of 10 R d⁻¹ and above, however, measurable radiation damage was evident; in particular, the growth rate of the irradiated fingerlings was significantly less than that of the controls. The apparent beneficial effect of irradiation at the lowest dose-rates mentioned above is a common feature of many irradiation effects experiments of all kinds. As yet there is no complete explanation of the phenomenon.

The genetic effects of radiation on aquatic organisms have also been studied, but so far these have been limited to a small range of species. In order to estimate the genetic effects at the population level one requires a knowledge of the mutation rate. The only data of value for aquatic organisms are those for the guppy (*Lebistes reticulatus*), for which values of 0.4 to $11 \times 10^{-3}\,\mathrm{Gy^{-1}}$ (0.4 to $11 \times 10^{-5}\,\mathrm{rad^{-1}}$) per gamete, and $2.5 \times 10^{-5}\,\mathrm{Gy^{-1}}$ ($2.5 \times 10^{-7}\,\mathrm{rad^{-1}}$) per locus have been determined. Using the assumption, based on DNA content of fish in general, that the number of loci coding for functional genes is of the order of 10^4, a mutation rate of $\sim 7 \times 10^{-2}\,\mathrm{Gy^{-1}}$ ($7 \times 10^{-4}\,\mathrm{rad^{-1}}$) per zygote seems reasonable. If, at the worst, all of these mutations were dominant lethals, less than one embryo in 1000 would be eliminated as the result of an integrated dose of 5 mGy (0.5 rad) by each of the parents. The study on plaice in the Irish Sea off Windscale mentioned earlier estimated that the fish could receive a life-time's dose (5 years – by which age most are caught) in the region of 90 mGy (9 rad).

In summing up the effects of radiation on the aquatic environment, the IAEA panel noted that the most sensitive organisms known to date are the eggs and embryos of teleost fish, for which some mortality has been observed in laboratory experiments at acute doses of the order of

1 Gy (100 rad). Chronic exposures have been observed to produce a number of minor metabolic effects at dose-rates of 10 mGy (1 rad) d^{-1}. It is evident that such dose-rates are well in excess of those given in table 6.9. But it should be noted that the scoring of effects is a reflection of the ability to detect them, and this is an area of research which is receiving increased attention. Nevertheless, it must be concluded that at the present rates of authorized discharges of radioactive waste to the aquatic environments – authorizations calculated to protect man, not the environment – the non-human biotic components are not being harmed in any way that is currently measurable, nor are they expected to be. Nor have those environments which are receiving such radioactive wastes as yet displayed any visible signs of 'harm'.

Finally, let us briefly return to the terrestrial environment. Acute radiation experiments have been made on entire terrestrial ecosystems using enormous point sources – up to 370 TBq (10kCi) ^{137}Cs. Pine forests irradiated with such sources for 200 h have demonstrated that dominant plant species are killed by exposures of about 3 kR and that growth is inhibited at exposures of about 100 R. Seasonal variations in radiation sensitivity have also been noted, but these differ from one ecosystem to another: for example, fields of short grass given seasonal exposures from a sealed ^{137}Cs source were more sensitive in late autumn and most resistant in the summer; studies with irradiated established fields, however, have shown them to be more sensitive in the spring and most resistant in the winter.

With regard to terrestrial vertebrates – particularly mammals – there are, of course, a vast amount of data which have been used for extrapolation to man. A number of experiments have been made to compare the effects of animals irradiated, and subsequently kept in the laboratory, with those irradiated and maintained in their natural environment. The results differ very little. Terrestrial invertebrates have also been studied, many of which, such as the insects, have been shown to be very resistant to irradiation. The details of such studies are not worth quoting here and it is sufficient to note that, as with the aquatic environment the doses, and dose rates, required to produce any observable effects are orders of magnitude greater than those obtaining in the environment as a result of discharges from nuclear power stations and reprocessing plants. Again it must be noted that this is an area of research which is continuing to develop and that improved techniques will undoubtedly show, experimentally, effects of irradiation at doses below those which are known to be effective at present. Whether these will ever be relevant to levels of radioactivity in the terrestrial environment is doubtful – particularly when discharges are, in most cases, so low that the resultant excess dose rates lie within the variation of the natural background radiation from one area to another.

7. The future

7.1. Introduction

The present state of energy production from nuclear power is very much an interim one. Above all there is the question of what can be done with regard to the safe disposal of the waste from spent nuclear fuel, be it reprocessed or left within the fuel rods. At present it is being stockpiled, in one form or another, wherever nuclear reactors are being operated around the world. Indeed it is becoming increasingly apparent that any large expansion of the nuclear power programme is likely to be dependent upon the ability to rid ourselves of this high-level waste in a manner which is safe both to man and to the environment.

There are other problems, although of a different scale. No commercial nuclear reactor has yet been de-commissioned and thus the question of what to do with the large bulk of radioactive materials which constitutes the reactor core and its associated servicing machinery has yet to be fully answered.

These are not the only areas to which attention is currently being directed. Consideration is being given to the future design of reactors. Mention was made in chapter 4 of the liquid-metal fast breeder reactor, and the possibility of using ^{232}Th instead of ^{238}U as a breeder material was also briefly discussed. Both are likely candidates for the future, particularly the former, but in addition to these the possibility of obtaining nuclear power from nuclear *fusion*, as opposed to *fission*, is also being actively pursued.

Finally, the acceptability of deliberately releasing radionuclides into the environment, one of the more intangible issues, requires some further discussion. In fact, it is this issue – as opposed to technological advances – which is as likely as any to determine the immediate future direction of nuclear power: it is therefore important that it is seen in perspective.

7.2. High-level radioactive waste disposal

The previous two chapters have been concerned with the low-level radioactive wastes deliberately introduced into the environment in a controlled manner: such wastes represent only a fraction of 1% of the total long-lived radioactivity arising from the nuclear fuel removed from reactors. As described in chapter 4 the reprocessing of spent fuel terminates, at present, with the production of a highly radioactive liquid which is stored on site in large, cooled, tanks. If the fuel is not reprocessed it is usually canned and stored in water-filled ponds. Both measures are clearly interim ones, and it is planned to dispose of this highly radioactive waste in such a way that it will never be a hazard to man nor, hopefully, to any other form of life. The only way such an assurance could be guaranteed would be to remove the radioactive waste from the planet altogether and this, as we shall see, has been considered.

Before considering where this waste should be put, however, it is first of all necessary to consider how one should handle it safely. Unreprocessed fuel rods obviously present no problem, except for their bulk, but the liquid reprocessed wastes present a greater difficulty. Studies have therefore been made over many years to devise satisfactory methods for the solidification of this waste. There are a number of properties required of a suitable material – a high resistance to thermal and radiation degradation, a high thermal conductivity, a high melting point, and a capacity to incorporate a relatively large quantity of the radioactive waste. The chief contender for such a material is a form of borosilicate glass. In the United Kingdom the mixture currently favoured is one of, on a weight basis, 25% radioactive waste, 43% silica (SiO_2) and 32% borax ($Na_2B_4O_7$). Other countries have also considered vitrification as a possibility – the French use a mixture of SiO_2, B_2O_3 and Al_2O_3 – but more recently attention has focused on the use of artificial rock, or 'synroc', as a more acceptable alternative. This consists of the production of crystalline minerals, related to known mineral types, which would incorporate the radioactive wastes into a fine-grained rock similar to basalt. The minerals chosen are those which are capable of dissolving different radionuclides, when collectively melted at ~1300° C, in such a manner that they become incorporated into the mineral crystals. For example, perovskite ($CaTiO_3$) has been shown to be very effective at incorporating rare earths and strontium; and another, zirconia (ZrO_2) is capable of incorporating long-lived actinide nuclides. Yet another line of research has centred on the possible incorporation of the waste into calcium aluminate cements. The conversion of the waste into calcine, a granular solid, by drying the waste at 800° C, or the production of crystalline–ceramic, or glass–ceramic bricks, are further possibilities.

Whichever process is used for solidification, however, one immediate advantage is the reduction in bulk. In the United Kingdom it has been estimated that its own installed nuclear generating capacity will rise to about 10 GW(e) by the year 1985, and to about 50 GW(e) by the year 2000, resulting in a cumulative 330 GW(e) years of generated electricity by that date. Such predictions may, or may not, be accurate but the interesting calculation is that the total activity of the wastes to be disposed of in the year 2000, following fuel reprocessing, would be approximately 1·11 EBq (30 MCi) actinide radionuclides and 111 EBq (3 GCi) fission product radionuclides; even after 1000 years this quantity will only have decayed to about 33·3 PBq (0·9 MCi) actinide, and 7·4 PBq (0·2 MCi) fission product, radionuclides. Feasibility studies have been made on incorporating the wastes into borosilicate glass which is then packaged into cylinders of stainless steel – or another suitable material – 3 m long and ½ m diameter. One such cylinder would contain the waste arising from about 5·5 tonne of spent fuel reduced to ~0·36 m^3 glass. Thus the waste arising from the entire programme would occupy no more than 2000 such cylinders.

Another, and more important, advantage in consolidating the liquid waste into some form of glass or other material is that it reduces the rate at which the radionuclides could enter the biosphere. A number of studies have been made to evaluate the degree of resistance to aqueous leaching of different materials, because it is always possible that water may eventually come into contact with them. Physical and mechanical integrity is also important because any disintegration of the solid will increase its surface area. Durability studies on borosilicate glass have been largely favourable, but so far such studies have been fairly short-term and it is not certain whether the high temperature and pressure regimes to which solidified waste may be exposed have been adequately simulated. Borosilicate glasses are made at a temperature of ~1000° C and some devitrification – a process involving the crystallization of certain constituents of the glass – will occur immediately on cooling. Further devitrification is to be expected over longer periods of time. The freshly made glass cylinders would, in any case, need to be cooled for about 10 years before disposal; and even after this period the inner temperature would still be nearly 200° C.

In addition to thermal stress, the glass would be subject to a radiation dose of about 0·3 MGy (30 Mrad) during the first 50 years, the major cause of damage being due to alpha decay. Apart from any large-scale fractures, however, such processes appear to have a very small effect on leaching rates. What does affect the leaching rate is the temperature of the leachant, and this may be an important consideration for ultimate disposal.

There are, therefore, a number of uncertainties which have yet to be resolved. Nevertheless there is little doubt that reprocessed fuel can, and will, be converted into a suitable solid form and packaged for disposal. Unreprocessed rods can also be packaged – a current Swedish programme is examining the possibilities of encapsulating fuel rods in highly impermeable artificial sapphire by means of a high-pressure technique. A more difficult problem is knowing where to put the packaged waste. For many years it has been assumed that the best option is that of deep geologic disposal. The criteria to be used in site selection have centred around the desirability of the site being remote from areas prone to earthquakes, devoid of surface water and far removed from rivers and lakes, devoid of minerals which are likely to be economically exploitable, and having favourable thermal conductive and physico-chemical properties – a rather daunting list. The most promising areas to meet these criteria have been salt formations, it having been assumed that their very presence is evidence of the total absence of water. But more recently it has been learned that the crystals in salt beds contain significant amounts of water as fluid inclusions, and water also occurs along intergranular boundaries. Moreover, at even fairly modest temperatures (150° C), this water can be released by the bursting of the inclusions, a process called decrepitation. Other geological structures are therefore being investigated, particularly hard rocks such as granite, and deep beds of clay. Because such rocks have been of little economical importance, little is known about their deeper geological structure.

Sufficient data are available, however, to enable some degree of mathematical modelling to be made of the likely consequences of the disposal of vitrified high-level waste in geological formations. The barriers which would prevent the return of the radioactivity to man fall into four main groups: the form of waste and its container; the containment of the waste within the geological formation; the degree of retardation of any released radioactivity as it migrates through the geological strata; and finally the dispersion and dilution of the radioactivity in the biosphere. With regard to the first barrier, it has already been stated that tests on different solidification techniques have been made for a number of years. Even making pessimistic assumptions that substances such as borosilicate glass will break up, the leaching rates are estimated to be very low. For the second barrier, geological containment, only probability estimates can be made. For instance the probability of a meteorite falling over the area, causing a release to the atmosphere of radioactive waste buried at a depth of 600 m, can be estimated at about 10^{-14} per year on the assumption that meteorite falls are randomly distributed over the surface of the Earth. Similarly, although not random, the probability of a new volcano erupting in an area of low tectonic plate activity such that it affects an area 1 km² is about 10^{-11} per year. Other natural events, resulting from long-term effects such as erosion, climatic changes – glaciation, pluvial episodes – and seismic events are unlikely to affect seriously the integrity of a particular waste disposal site, but could allow water to enter and thus increase the rate of radionuclide release. Human action could also breach the geological containment barrier by direct

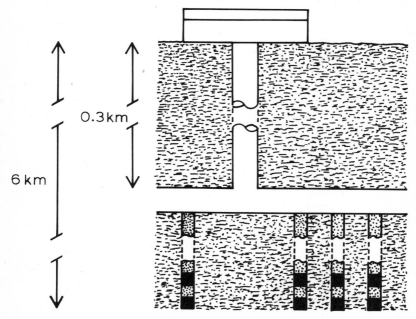

Figure 7.1. Solid-waste disposal concept in which vitrified high-level waste, in canisters, is buried in a matrix of holes on the floor of a mined facility. Adapted from Hill, M. D. and Grimwood, P. D. *Preliminary Assessment of the Radiological Aspects of Disposal of High-Level Waste in Geologic Formations*, NRPB-R69; coutesy of the authors and of the National Radiological Protection Board.

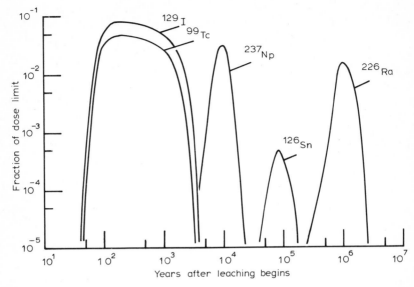

Figure 7.2. Pessimistic theoretical calculations of the fraction of the dose limit to individuals of the general public from drinking water which has become contaminated as a result of groundwater leaking from canisters buried as in figure 7.1. The calculations assume that all [129]I present in reprocessed fuel is retained for high-level waste disposal. Adapted (with additional data kindly supplied by the authors) from Hill, M. D. and Grimwood, P. D. *Preliminary Assessment of the Radiological Aspects of Disposal of High-Level Waste in Geologic Formations*, NRPB-R69; courtesy of the authors and of the National Radiological Protection Board.

drilling and mining, although such actions are unlikely to result in widespread contamination. The waste would certainly be buried at depths beyond the effect of nuclear weapon explosions on the surface.

Should both the container and the geological containment fail, the third barrier is the retardation of the radioactivity as it passes through the geological medium itself. The major processes to be considered are those of dispersion, convection, adsorption, and radioactive decay of radionuclides travelling with any ground-water which may have entered the breached canisters. Finally there is the dilution within the biosphere. As a pathway back to man this is likely to be of secondary importance; if it enters a body of freshwater the most important route of exposure would probably be the consumption of contaminated drinking water.

Using all of the data currently available, a mathematical model – one of many – has been derived to estimate the consequences of burying the canisters, theoretically arising in the United Kingdom by the year 2000, in a hard rock repository.† The general layout of the presumed disposal site is that of figure 7.1, in which the canisters are placed in a matrix of drilled holes within the floor of a subterranean cavern. Assuming that ground-water enters the containers 1000 years after burial, the predicted peaks of annual individual doses to members of the general public after leaching begins – from contaminated drinking water – are shown in figure 7.2. Four radionuclides were considered to be of the greatest importance – ^{99}Tc, ^{129}I, ^{226}Ra and ^{237}Np; although none would result in an exposure greater than 10% of that currently recommended by the ICRP for individual members of the general public.

Next to geological disposal the most promising disposal site is the sea bed: not the relatively shallow waters which receive low-level radioactive waste, but the floors of the oceans. The sea covers 70% of the earth's surface and is divided into three principal regions, each occupying approximately a third of the total area. The *continental margins*, which include the continental shelves, inland seas, marginal plateaus and so on are clearly unsuitable. Equally unsuitable are the seismically active *mid-oceanic ridges*, the spreading centres of the sea floor which form the 'constructive' edges of the Earth's tectonic plates. Between these two areas is the *ocean basin floor* which comprises the flat abyssal plains, the gently rolling abyssal hills, and the deep-ocean trenches. Of these three the regions of abyssal hills are most favoured because they are covered in places with 50 to 100 m clay which overlie a more consolidated layer of lithified sediment. Both of these layers lie on top of a layer of basaltic rock. The abyssal hills – or mid-plate/mid-gyre regions, as they are also called – have the added qualities of being seismically passive, relatively stable, relatively unproductive biologically and limited in resources as far as can be judged at present.

Two possibilities can be envisaged: disposal *on* the sea floor or disposal *into* the sea floor. The latter possibility is not dissimilar from the terrestrial geological option, but has the additional barrier to the transfer of radionuclides back to man of the enormity of the sea itself. If the packages were to be dumped directly onto the sea bed they would at least be cooled, so that problems arising from self-heating would be minimized. Shielding of the packages additional to that needed for land burial would be required, particularly in order to contain the shorter-lived fission product radionuclides. This would

† National Radiological Protection Board, 1978, NRPB-R69.

certainly be no problem for periods of ~100 years but a mean containment period of ~1000 years is more desirable. A number of alloys are considered to have sufficient sea water corrosion-resistant properties. Disposal into the sea bed is clearly a more satisfactory method but less economical and technically more difficult. Where areas of the sea bed have a deep layer of sediment, free-fall containers (figure 7.3) could be used. These would be streamlined, finned, con-

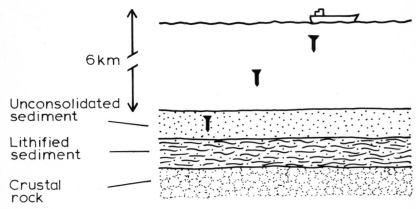

Figure 7.3. Possible concept for a free-fall disposal of canisters, containing high-level waste, into the ocean basin floor.

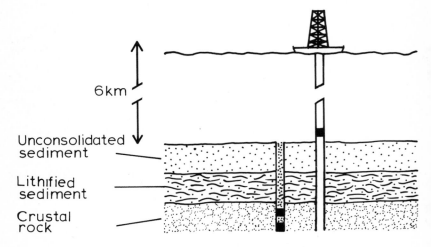

Figure 7.4. Possible concept for the emplacement, by drilling and subsequent backfilling, of canisters containing high-level waste into the ocean basin floor.

tainers and a small one – about 5 m long – could be expected to penetrate some 10 m or more into the sea bed when released in water of 6 km depth. A more satisfactory method (figure 7.4) would be to drill holes into the sea bed, place the canisters within the shafts, and backfill. An additional barrier would therefore exist in the form of the depth of sea-bed sediment lying above the canisters.

Apart from the technical difficulties involved in sea-bed burial the greatest problem at present in evaluating the sea-bed option – with or without burial – is our considerable ignorance of the physical, chemical and biological characteristics of the deep oceans. It is generally believed that the rates of turnover of the deep ocean waters are very slow, of the order of hundreds of years, but these are known very imprecisely. The role of large climatic changes in effecting deep ocean water movements are similarly largely unknown. On the shorter time scale, data are lacking on both horizontal and vertical diffusion coefficients at depths greater than 1 km, and research programmes made over the next few years will largely be addressed to the physical inadequacies of our knowledge of the deep oceans.

It is also true to say that we know almost nothing about the biology of the deep oceans, although most of the data available indicate quite a variety of animal phyla, with polychaetes, lamellibranch molluscs, crustaceans, and the ubiquitous echinoderms dominating the benthic fauna. The use of baited traps has shown that the abundance of mobile animals is probably greater than previously thought; but it is still extremely difficult to make sensible relative abundance estimates with the techniques currently used. Perhaps the overriding consideration is that although apparently hostile – near freezing temperatures, enormous pressures, total absence of light and low food levels – it is nevertheless a very stable environment. It is therefore assumed that deep-sea communities are likely to be less able to adapt to environmental changes than others, and thus sensitive to relatively small perturbations. There is no reason to suspect that they would be any more sensitive to irradiation than any other community, however; the natural radiation regime in which they evolved is similar to that of coastal waters, and it is in any case necessary to consider the enormity of the total areas of the ocean basin floor relative to the comparatively minute areas where any waste would be placed.

Another obstacle which would need to be overcome is that deriving from international opinion. Land disposal of high-level waste is at least a national option whereas deep sea disposal would require some degree of international agreement. From the scientific viewpoint the best that can be attempted is an evaluation of the likely consequences of the radioactivity being released from the containers. Mathematical models have again been used in this context. An earlier study† on the disposal of the theoretical quantities of waste arising in the United Kingdom by the year 2000 considered the consequences of its disposal on to the floor of the North Atlantic Ocean. The quantities considered were somewhat larger – 33·3 EBq (0·9 GCi) actinides and 3700 EBq (100 GCi) fission products – than in the land burial study. Again the waste was envisaged as being incorporated into borosilicate glass, and it was pessimistically assumed that the glass came into contact with the sea water immediately after being dumped onto the sea bed. Exposure to man in this case would arise almost entirely from the ingestion of contaminated marine food species. The greatest fraction of annual intake relative to that recommended by the ICRP for individual members of the public was estimated by this model to be no greater than about 1%, after a period of 50 to 100 years, as a result of ingesting either deep sea fish species, or plankton – both rather unlikely food chains direct to man. A more realistic assumption was also made; that the principal route back to man would be via the consumption of near-surface water fish. This would result in the highest individual dose to man being only about 0·04% of ICRP-recommended levels – 50 years after disposal of the waste – as a result of ^{90}Sr and ^{137}Cs in the fish. All other individual exposures to the radionuclides would result in less than 0·001% of ICRP-recommended limits, regardless of the time scale. In terms of collective dose, again the highest values arise from ^{90}Sr and ^{137}Cs; but other radionuclides such as ^{241}Am, ^{243}Am, ^{210}Pb, ^{226}Ra and ^{126}Sn become increasingly important after 100 to 1000 years.

These values, which assume the worst imaginable consequences, initially appear to be fairly acceptable as far as man is concerned. As an option for the future, therefore, disposal into the deep oceans will continue to be seriously considered. Improved mathematical models have been developed to describe the physical processes of radionuclide dispersion and diffusion in the oceans, and as more data become available to refine such models a better comparison of land and sea disposal options can be made. It should also be noted that the models which have been made to date, with regard to all of these different possibilities of high-level waste disposal, have deliberately attempted to overestimate the dose to man to test whether doses to individuals would exceed the ICRP recommended limits. In view of the current philosophy of optimization, as discussed in chapters 5 and 6, it is also necessary to produce models which reflect the most probable collec-

† National Radiological Protection Board, 1976, NRPB-R48.

tive dose, or collective dose commitment, as used in figure 5.1. Unfortunately, there are insufficient data available to produce such models, and this is currently an area in which much research is being done. Data are particularly lacking with regard to the realistic modelling of the sea bed disposal options.

Other disposal schemes have also been proposed. One of the most unacceptable involves disposal into the Antarctic ice sheets, although one can immediately appreciate the simplicity of the scheme. The Antarctic continent is covered in a sheet of ice up to 4 km deep. If one assumed an average ice thickness of 3 km, a canister introduced into a hole on the surface would take about 5 to 10 years to melt its way down to the underlying bedrock, the ice reforming over it as it descended. The descent could be slowed down by attaching cables, 200 to 500 m long, to the surface. These would then hold the canister at a predetermined depth from which it could be retrieved if necessary. But the fate of the ice sheet itself over such periods of time, dependent as it would be on global climatic conditions, cannot be predicted and therefore completely negates this option – apart from the undoubted international objections.

The most satisfactory place to dispose of the high-active waste is the sun. The possibility of using the space shuttle in this context has been examined, the shuttle carrying in its hold a package of waste containers plus a propelling device which would tow it to the sun. Apart from the enormous expense involved in such an operation – even though minimized by using a space shuttle instead of non-retrievable rockets – the possibility of an accident at launch, or before reaching orbit, largely rules out this option, even though it is technically feasible.

Finally, there is one further possibility under study, a possibility markedly different from those discussed above, and that is the transmutation of the long-lived actinide radionuclides within a reactor. The basic idea is to remove the actinide waste from the high-level waste and re-introduce it into a reactor core. There, by successive neutron capture, different nuclides would be created, some of which would undergo fission and thus give rise to the shorter-lived fission product radionuclides. Feasibility studies are currently being made to determine what balance can be struck between neutron capture which induces fission of a nucleus, and neutron capture which simply produces heavier nuclides which may be more hazardous than the initial ones. Thus calculations have to be made on the effect of different neutron fluxes on the chances of neutron capture, on the chances of inducing fission, and more importantly on the *ratio* of these two possibilities occurring. It seems most likely that only fast breeder reactor conditions will be useful in this respect and, as such reactors have not yet been built on a commercial scale, the best options for disposal of the high-level waste remain those of geological disposal or emplacement in or on the sea bed.

7.3.   Decommissioning nuclear reactors

Nuclear reactors will wear out and also become technically obsolete. Deterioration is to be expected in the installed electrical and mechanical equipment and eventually the decision will be made to shut a reactor down permanently. The majority of reactors currently in use have been built with a life expectancy of at least 25 years and thus none of the larger commercial reactors has yet been decommissioned. A number of small experimental reactors have been decommissioned, however, and thus some experience has been gained. Repairs which have necessarily been made during the life of reactors have also yielded much useful experience. Nevertheless, the reality of having to dismantle and dispose of a number of large nuclear reactors has yet to be faced.

The full extent to which any particular site will be completely dismantled will depend on the future use to which it is to be put. In many countries the sites on which nuclear power stations have been built have been carefully chosen on the basis of a number of criteria, and there are therefore strong incentives to use the same site, if possible, for a replacement reactor. Initially the fuel will be removed and thus the only subsequent source of radioactivity will be the longer lived neutron activation products formed within the reactor materials. Of these, ^{60}Co is considered to be the greatest single hazard, being a high-energy gamma emitter. For a twin magnox reactor station of about 500 MW(e) capacity it has been estimated that 1 year after shutdown the total reactor inventory of radioactivity would be about 126 PBq (3·4 MCi).

There are three possible options remaining after the fuel has been removed. If the site was still in use, for example because of a replacement reactor being built alongside, it would be possible simply to mothball the reactor for an extended period of time. During this period it would require continued surveillance to a degree that would only be feasible if trained personnel were already on site. A more realistic option for sites which are not intended for further use is one of long-term entombment. This would involve the removal of all equipment and ancillary structures which could be fairly easily dismantled, followed by the sealing of the reactor itself to prevent any form of access. For either of these two options, however, it must be assumed that ultimately the reactor will have to be completely dismantled. A delay of at least 50 years would be required to derive any real benefit from the decay of the ^{60}Co; although there would be little advantage in waiting for more than 100 years. Even when complete decommissioning is to be undertaken the process is estimated to take about 10 years.

The reactor would need to be dismantled from the inside outwards, in order to take full advantage of the biological shield. After removal of all peripheral structures the core of a magnox reactor would have to be removed graphite block by graphite block. Next the pressure vessel itself, and all its associated components, would need to be removed, and finally the inner layers of the concrete biological shield. The remainder of the outside shell could then be dismantled.

Although not highly active, the structure of a reactor core represents a considerable waste disposal problem because of its sheer bulk – over 15 000 tonne. Transport alone would be a major effort if a high degree of shielding was necessary. As to where the materials should be placed, again the only real choice is between some form of land burial or disposal into the deep sea. For land burial, consideration may be given in the future to the siting and construction of the reactor core below ground level so that once the above-ground structures had been cleared away the core itself could be simply sealed in.

7.4. *Nuclear fusion reactors*

One of the possibilities for energy production in the future is that of nuclear fusion. Whereas our present source of nuclear power is derived from the energy liberated as a result of the fission of very heavy nuclei, power from nuclear *fusion* would be derived from the energy released when two light nuclei are fused together. This is possible because the sum of the masses of individual light nuclei is greater than the mass of the nucleus formed if the two nuclei are combined. As with nuclear fission, it is necessary for the reaction to be self-sustaining – and thus more energy needs to be released than is consumed in initiating the reaction – if it is to be considered as a source of power.

For two atomic nuclei to interact they must possess enough kinetic energy to overcome the electrostatic repulsion barrier which keeps them apart. The energy required for nuclei of hydrogen, the lightest element, to react together is about 0·1 MeV. This energy can be attained in a particle accelerator by directly imparting kinetic energy; it could also be attained, in principal, by increasing the temperature but the temperatures required are enormous – of the order of 10^9 C – if the *average* energy of the particles is to be 0·1 MeV. At these very high temperatures gases do not exist as molecules but as completely ionized systems of positively charged nuclei and negatively charged electrons – and thus electrically neutral as a whole. Such systems are called *plasmas*. Within the plasma not *all* of the nuclei will have energies close to the average *all* of the time. Similarly, although the energy required for nuclei of hydrogen to fuse is about 0·1 MeV, there is a *probability* that fusion could take place at all temperatures. Thus, because the chances of fusion taking place is the product of these two different probabilities, it is theoretically possible to obtain a desired level of fusion at temperatures below that required for 0·1 MeV average particle energies to be sustained.

The reaction which is considered to be the most favourable for nuclear fusion is that between two isotopes of hydrogen, tritium and deuterium, i.e.

$$^3H + {}^2H \rightarrow {}^4He + 1\,n + 17 \cdot 6\,MeV \qquad (7.1)$$

The temperature at which the rate of energy production in this reaction exceeds the rate of energy loss – by bremsstrahlung – is about 5×10^{7} ° C in an ideal system.

If fusion is to be considered as a source of power it is also important to realize that the total, useful, energy recovered must be at least sufficient to maintain the temperature of the reaction. This criterion is expressed in terms of the number of reacting nuclei per unit volume, and the time during which the reaction takes place. There are, therefore, three basic requirements for the construction of a fusion reactor as a source of power: an optimum number of nuclei per unit volume – about 10^{14} cm^{-3}; a temperature in excess of 10^{8}° C; and the sustainment of these conditions for periods of several seconds. Sustaining, and thus confining, the reaction is one of the biggest difficulties. At present the only known means of confining a high-temperature plasma is the use of magnetic fields. These confine the plasma because it is difficult for charged particles to cross them although there is, nevertheless, a partial 'mixing' between the plasma and the magnetic fields and this causes the particles to spiral along the lines of force.

There have been a number of different approaches towards plasma confinement. One of the earliest was that of the *pinch effect* in which a very strong current is passed through a gas such as deuterium at very low pressure. The electric current both heats the gas and provides an encircling net of magnetic lines of force which compresses the gas into the centre of the vessel – usually in the shape of a thick ring, or *toroid* as it is called. Unfortunately, the plasma in such systems could only be maintained for a few microseconds. It was subsequently learned that plasma stability could be improved by the addition of longitudinal magnetic fields (figure 7.5), a development which culminated in the production of the *tokamak* system, pioneered by the USSR. Tokamak is a word derived from *tok*, meaning electric current, and *mag*, for magnetic. By 1977, tokamak systems had, in various laboratories in a number of countries, succeeded in attaining plasma densities in the range of 10^{13} cm^{-3}, temperatures of about $10^{7\circ}$ C, and sustainment times of about 1 s. A source of heating other than that derived from the toroidal coils has been required to attain these conditions.

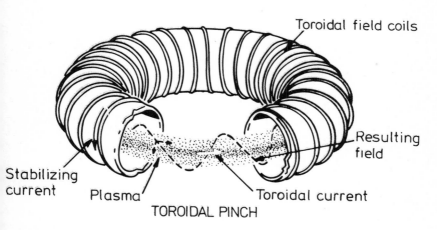

Figure 7.5.   Schematic diagram of a tokamak toroidal pinch. From Pease, R. S., 1978, *Nuclear Energy*, **17**, 271; courtesy of the author and The British Nuclear Energy Society.

Other confinement systems have been developed in which a helical arrangement of magnetic field coils, with current flowing in opposite directions (figure 7.6), lies within a toroidal field: such systems are called *stellerators*. Temperatures and plasma densities similar to those of tokamaks have been achieved.

A more recent approach to controlled fusion is one based on an entirely different concept – *inertial confinement* – in which high-powered pulsed laser beams are used to create small-scale explosions of deuterium/tritium pellets. The fusion reaction is induced by compressing the pellets to densities which greatly exceed their normal,

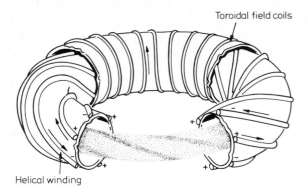

Figure 7.6.   Schematic diagram of the magnetic field coils in a stellarator. From Pease, R. S., 1978, *Nuclear Energy*, **17**, 271; courtesy of the author and The British Nuclear Energy Society.

solid-state, densities. Glass microspheres filled with a mixture of deuterium and tritium gases have been used to attain temperatures in excess of $10^{7\circ}$ C, and further developments are expected soon in this rapidly developing line of research.

Encouraging as all of these results are, the fusion reactor is still, however, largely theoretical. There seems to be little doubt that the basic physical requirements of a maintained fusion reaction will soon be achieved, but the step from laboratory experimentation to commercial fusion reactor is a very large one indeed. A number of design studies have been made and these have highlighted one of the most challenging areas, that of reactor service and maintenance. Very high demands are made of the inner lining of a toroidal system. Not only is it required to maintain a high vacuum and high thermal gradients, but it is also subjected to a high neutron flux resulting from the fusion reaction and to direct damage when the plasma becomes unstable and touches the vessel wall. It is therefore envisaged that this inner liner will have to be frequently mended or replaced. In addition to the complex array of superconductors for magnetic fields around the torus, there will also be a blanket of lithium – in the form of LiO_2 or $LiAlO_2$ – in order to breed ^3H from the reaction

$$^6Li + n \rightarrow \,^3H + \,^4He$$

The tritium would be used as fuel, together with deuterium. Yet a further complication is the means by which the cooling gas – for which He is preferred – will actually remove the heat generated, and the means by which the 'exhaust' plasma will be extracted and replaced.

In view of these difficulties, it may well be asked what is there to be gained? There are two theoretical answers. One is that the supply of fuel, deuterium and tritium derived from hydrogen and lithium, respectively, would be virtually limitless. Secondly, the major form of waste products would be tritium, which has the short half-life of 12·4 years and for which virtually limitless capacity for dilution exists in the form of the world's oceans. As a source of energy supply, however, it is unlikely that fusion will make any significant contribution until well into the next century.

7.5. The acceptability of continued discharges into the environment

Although the vast majority of informed scientific opinion accepts that the quantities of radioactivity which have been discharged into the environment, and which are currently being discharged, may be regarded as 'safe', important areas of controversy still exist; they possibly always will. There are at present, however, a number of areas which are of sufficient concern for their evaluation to be of some importance to the expansion of nuclear power programmes in many countries.

One such area of controversy is that of the recommended values of exposure considered as being acceptable for the general public. The most recent recommendations of the ICRP, presented in ICRP 26 (1977), are already being challenged. As discussed in chapter 3 these recommendations have relied heavily, although not exclusively, on data derived from the survivors of Hiroshima and Nagasaki. With regard to the risk factors for some irradiation-induced cancers, it has been argued that data from other sources – uranium miners, patients irradiated as a result of various medical practices, populations accidentally irradiated in nuclear weapons tests such as those in the Marshall Islands – all yield higher risk factors than the Japanese survivors. The types of cancer for which this appears to have some validity are those of the thyroid, breast and lung, the risk factors being some five or six times as great as those derived from Hiroshima and Nagaski. For leukaemia the derived risk factor is the same, but it has been pointed out that even this similarity throws some doubt on their derivation because the Japanese data were based on whole-body dose calculations whereas the data from the other sources have been based on the assumption that only part of the bone marrow was exposed. It has therefore been suggested that the current recommendations of the ICRP may be too lenient by a factor of about five, possible more.

It is interesting to reflect on the degree to which recommended acceptable dose levels have continuously decreased over the last 25 years in some countries, particularly in the USA. The value recommended for the public by the US National Council on Radiation Protection and Measurements (NCRP) in 1952 was 15 mSv y^{-1} (1·5 rem y^{-1}) for any body organ. In 1959 this was reduced to 5 mSv y^{-1} (0·5 rem y^{-1}) as a result of the recommendations of the ICRP – a value still recommended by them in 1977 (ICRP No. 26). However, the value suggested in 1974 by the US Energy Research and Development Agency (ERDA) for persons living near a PWR nuclear power plant was 0·05 mSv y^{-1} (5 mrem y^{-1}). This is much lower than the dose received from the natural radioactivity within one's own body!

As stated in chapter 5, the application of an annual dose equivalent limit of 5 mSv (0·5 rem) is likely to result in average annual dose equivalents of less than 0·5 mSv (50 mrem). Assuming a total risk of about 10^{-2} Sv^{-1} (10^{-4} rem^{-1}), a life-time's exposure should ideally correspond to 1 mSv y^{-1} (0·1 rem y^{-1}). In fact the ICRP suggests that in rare cases, where the doses to a few individuals of the general public are actually found to be received at high rates over prolonged periods, it would be prudent to take measures to restrict their dose to 1 mSv y^{-1} (0·1 rem y^{-1}).

Unfortunately the use of dose limits, indeed the use of the word 'limit', is frequently misunderstood by the public at large. Fears are often expressed that 'the authorities' may not always get their sums right, that environmental surveillance may not always be sufficiently vigorous, that some practice may continue unnoticed and so on; in short, that something might happen which would result in 'the limits' being exceeded, with dire consequences. It is quite possible that occasionally an individual may exceed the recommended annual dose equivalent limit, but this will not result in any deleterious effect; it merely makes the risk, the chances, of a subsequent deleterious effect as a result of that exposure very, very, slightly higher.

It should also be recalled that any calculation on risk is taken together with assessments of benefit, and of cost. In fact it is worth restating the three main features of the ICRP recommendations of dose limitation: (a) that *no* practice shall be adopted unless its introduction produces a positive net benefit; (b) that *all* exposures shall be kept as low as reasonably achievable, economic and social factors being taken into account, and (c) that the dose equivalent to individuals shall *not* exceed the limits recommended for the appropriate circumstances by the Commission. The limits of dose equivalents which shall not be exceeded for the public represent a level of fatality risk which is of the order of 10^{-6} to 10^{-5} per year, according to the current views of ICRP, and these limits, and thus risks, are to be even further reduced on an optimization basis.

It is worth briefly considering the levels of risk at which public money may be spent in order to reduce them. In a report of the Royal Commission on Environmental Pollution (UK)† it was suggested that fatality risks which were greater than 10^{-3} per year would normally be considered unacceptable in society. (Smoking 10 cigarettes a day falls within this category, however, with a risk to the individual of about 1 in 400 per year). With risks of about 10^{-4} it was suggested that public money would normally be spent to try to eliminate the causes, and to mitigate their effects. Traffic accidents fall within this category. Risks which are below 10^{-5} were considered to be individual risks which might warrant specific warnings being given under certain circumstances, but that risks which were in the region of 10^{-6} per year were generally accepted by the public without concern. Which is just as well, because the risk of being struck by lightning is about 0.5×10^{-6} per year! Of more relevance, however is a comparison with motor transport, for which most people would agree there is usually an obvious benefit, and risk, to be taken into account. A risk rate of 10^{-6} per year is equivalent to the risk involved in driving 50 miles in the United Kingdom. Incidentally, as mentioned briefly in chapter 3, the risk of developing leukaemia (in the United Kingdom) as a result of natural causes – whatever *they* may be – is about 0.5×10^{-4} per year.

Notwithstanding all of these arguments, the continuing concern over what constitutes an acceptable dose limit for exposed workers, or for the public, relates to that which was implied at the beginning of this section: our lack of good data on the effects of radiation, particularly on man, at low levels of exposure. One approach to fill this gap in our knowledge has been the use of epidemiological studies on persons who have worked in certain branches of the nuclear industries, or who have otherwise been occupationally exposed to ionizing radiations. The occurrences of different forms of cancer have been recorded, and these data have then been related to estimates of absorbed dose, or compared with the rates of cancer induction in the public at large. This would at first appear to be a very sound procedure. Unfortunately the number of persons concerned is fairly small, and the rates of induction of many of the cancers are fortunately rather low. Inevitably, therefore, the application of different statistical techniques to such data produce different, and often conflicting, conclusions. In view of the fact that estimates of very low absorbed dose rates are difficult to make because of the many variables involved – for both the public, and exposed workers, the largest fraction of absorbed radiation, over

† Royal Commission on Environmental Pollution, 1976, Sixth Report, Cmnd 6618; Chairman Sir Brian Flowers, (London: HMSO).

and above a variable background rate, arises from medical care – and bearing in mind that most neoplastic diseases are not only 'naturally' occurring but may also be induced by other carcinogens, it is hardly surprising that each study, and its interpretation, produces different results. There is also the difficulty that, in any attempt at deductive analysis, a correlation, no matter how statistically valid, does not necessarily imply cause and effect. It seems more likely, therefore, that although epidemiological studies may well be used in arguments for and against particular dose limits, the only data which can be obtained scientifically on the effects of low-level exposure will be those stemming from repeatable laboratory studies using improved biological techniques and methods of dosimetry.

An equally difficult area of controversy centres around the philosophy of introducing very long-lived radionuclides from current reprocessing activities, and ultimately in the form of high-level waste, into the environment. The long-lived radionuclides present a challenge in a number of ways. Very little is known of the behaviour of some of them. The transuranium elements, for instance, being largely man-made, are new to the environment and thus as yet we have a very incomplete picture of their long-term behaviour. This is also true of elements such as technetium, which has been very little studied. Undoubtedly changes may well be made with regard to the recommended discharge rates of a number of such nuclides on the basis of new data, either with regard to their metabolism by man or their behaviour in the environment. But again it is important that such reservations are not seen out of all perspective. This is not to state a case for, or against, the use of nuclear power. The advocacy of pursuing nuclear power as a means of producing energy is a topic far beyond the scope of this book, and one which involves a very large number of far reaching questions. With regard to radioactive waste management, however, one cannot ignore the fact that our current industrial practices result in the discharge of innumerable chemicals into the environment. Amongst these are known toxic elements – such as mercury, cadmium, lead and arsenic – which are potentially toxic for ever, for they do not decay. Regulations are gradually being drawn up with regard to the deliberate discharge of other chemicals into the environment by industries, and these are likely to undergo far greater modification than those relating to the discharge of radioactivity. Indeed there are no other sets of standards, or codes of practice, which even begin to compare with those which have already been set for the nuclear industries.

Appendix 1. Prefixes for SI units

Prefix	Symbol	Factor
exa	E	10^{18}
peta	P	10^{15}
tera	T	10^{12}
giga	G	10^{9}
mega	M	10^{6}
kilo	k	10^{3}
milli	m	10^{-3}
micro	μ	10^{-6}
nano	n	10^{-9}
pico	p	10^{-12}
femto	f	10^{-15}
atto	a	10^{-18}

Appendix 2. Some radionuclides and their half-lives

Element	Radionuclide	Half-life		Principal radiations	
Actinium	^{228}Ac	6	h	β^-	γ
Americium	^{241}Am	433	y	α	γ
	^{243}Am	$7 \cdot 9 \times 10^3$	y	α	γ
Antimony	^{124}Sb	$60 \cdot 2$	d	β^-	γ
	^{125}Sb	$2 \cdot 8$	y	β^-	γ
Argon	^{41}Ar	$1 \cdot 8$	h	β^-	γ
Barium	^{137m}Ba	$2 \cdot 6$	m	γ	
	^{139}Ba	$1 \cdot 4$	h	β^-	γ
	^{140}Ba	$12 \cdot 8$	d	β^-	γ
Beryllium	^{7}Be	$53 \cdot 3$	d	γ	
Bismuth	^{210}Bi	$5 \cdot 0$	d	β^-	
	^{214}Bi	$19 \cdot 8$	m	β^-	γ
Bromine	^{87}Br	55	s	β^-	γ
Caesium	^{134}Cs	$2 \cdot 1$	y	β^-	γ
	^{137}Cs	$30 \cdot 1$	y	β^-	
	^{139}Cs	$9 \cdot 5$	m	β^-	γ
Calcium	^{45}Ca	164	d	β^-	
Carbon	^{14}C	$5 \cdot 7 \times 10^3$	y	β^-	
Cerium	^{141}Ce	33	d	β^-	γ
	^{144}Ce	284	d	β^-	γ
Chromium	^{51}Cr	$27 \cdot 7$	d	γ	
Cobalt	^{57}Co	271	d	γ	
	^{58}Co	$70 \cdot 8$	d	β^+	γ
	^{60}Co	$5 \cdot 3$	y	β^-	γ
Hydrogen	^{3}H	$12 \cdot 4$	y	β^-	
Iodine	^{129}I	$1 \cdot 6 \times 10^7$	y	β^-	γ
	^{131}I	$8 \cdot 1$	d	β^-	γ
	^{132}I	$2 \cdot 3$	h	β^-	γ

Element	Radionuclide	Half-life		Principal radiations	
Iron	^{55}Fe	2·7	y	X-rays	
	^{59}Fe	44·6	d	β^-	γ
Krypton	^{85}Kr	10·7	y	β^-	γ
	^{87}Kr	1·3	h	β^-	γ
Lanthanum	^{140}La	1·7	d	β^-	γ
Lead	^{210}Pb	22·3	y	β^-	
	^{214}Pb	26·8	m	β^-	γ
Manganese	^{54}Mn	313	d	γ	
Molybdenum	^{99}Mo	1·1	d	β^-	γ
Neodymium	^{147}Nd	11	d	β^-	γ
Neptunium	^{237}Np	$2·1 \times 10^6$	y	α	γ
	^{238}Np	2·1	d	β^-	γ
Nitrogen	^{16}N	7·2	s	β^-	γ
Niobium	^{95}Nb	35·1	d	β^-	γ
Phosphorus	^{32}P	14·3	d	β^-	
Plutonium	^{237}Pu	46	d	γ	
	^{238}Pu	87·8	y	α	
	^{239}Pu	$2·4 \times 10^4$	y	α	
	^{240}Pu	$6·5 \times 10^3$	y	α	
Polonium	^{210}Po	138	d	α	
Potassium	^{40}K	$1·3 \times 10^9$	y	β^-	γ
Praseodymium	^{143}Pr	13·6	d	β^-	
	^{144}Pr	17	m	β^-	γ
Promethium	^{147}Pm	2·6	y	β^-	
Protactinium	^{233}Pa	27	d	β^-	γ
	^{234m}Pa	1·2	m	β^-	
Radium	^{224}Ra	3·6	d	α	γ
	^{226}Ra	$1·6 \times 10^3$	y	α	γ
	^{228}Ra	5·8	y	β^-	
Radon	^{220}Rn	55	s	α	
	^{222}Rn	3·8	d	α	
Rhodium	^{103m}Rh	57	m	X-rays	
	^{106}Rh	30	s	β^-	γ
Rubidium	^{87}Rb	$4·8 \times 10^{10}$	y	β^-	
Ruthenium	^{103}Ru	39·5	d	β^-	γ
	^{106}Ru	1·0	y	β^-	
Silver	^{110m}Ag	253	d	β^-	γ
Sodium	^{22}Na	2·6	y	β^+	γ
	^{24}Na	15	h	β^-	γ
Strontium	^{89}Sr	50·5	d	β^-	
	^{90}Sr	28·5	y	β^-	
Sulphur	^{35}S	87·4	d	β^-	
Technetium	^{99}Tc	$2·1 \times 10^5$	y	β^-	
Tellurium	^{132}Te	1·3	h	β^-	γ
	^{135}Te	2	m	β^-	γ

Element	Radionuclide	Half-life		Principal radiations	
Thallium	^{208}Tl	3·1	m	β^-	γ
Thorium	^{228}Th	1·9	y	α	
	^{230}Th	$7 \cdot 7 \times 10^4$	y	α	
	^{232}Th	$1 \cdot 4 \times 10^{10}$	y	α	
	^{233}Th	22·2	m	β^-	γ
	^{234}Th	24·1	d	β^-	γ
Tin	^{126}Sn	10^5	y	β^-	γ
Uranium	^{233}U	$1 \cdot 6 \times 10^5$	y	α	
	^{234}U	$2 \cdot 5 \times 10^5$	y	α	
	^{235}U	$7 \cdot 1 \times 10^8$	y	α	γ
	^{236}U	$2 \cdot 4 \times 10^7$	y	α	
	^{238}U	$4 \cdot 5 \times 10^9$	y	α	
	^{239}U	23·5	m	β^-	γ
Xenon	^{133}Xe	5·3	d	β^-	γ
	^{135}Xe	9·2	h	β^-	γ
	^{139}Xe	43	s	β^-	γ
Yttrium	^{90}Y	2·7	d	β^-	
	^{91}Y	58·5	d	β^-	
	^{95}Y	11	m	β^-	γ
Zinc	^{65}Zn	244	d	β^+	γ
Zirconium	^{95}Zr	64	d	β^-	γ

y = years; d = days; h = hours; m = minutes; s = seconds.

Appendix 3. Serial decay

When a radioactive parent decays to a radioactive daughter, in a simple parent–daughter relationship, the number of atoms of the daughter nuclide (N_B) is

$$N_B = \frac{\lambda_A}{\lambda_B - \lambda_A} \, N^0{}_A(e^{-\lambda_A t} - e^{\lambda_B t}) + N^0{}_B e^{-\lambda_B t} \qquad (A3.1)$$

where $N^0{}_A$ and $N^0{}_B$ are the number of atoms of parent and daughter, respectively, at $t = 0$. Thus where radionuclide [A], decays to [B] which in turn decays to a stable nuclide [C]

$$[A] \xrightarrow{\lambda_A} [B] \xrightarrow{\lambda_B} [C]$$

the relative amounts of [A] and [B] present depend upon their respective decay constants, λ_A and λ_B. If the half-life of [A] is very much greater than that of [B], then the quantities of activity, A_A and A_B, present can be described by

$$A_B = A_A(1 - e^{-\lambda_B t}) \qquad (A3.2)$$

This is known as *secular equilibrium* and is described diagrammatically in figure A3.1. An example is the decay of ${}^{90}Sr$ to ${}^{90}Y$.

For equation (A3.2) the quantity of parent radionuclide remains substantially constant. Two other general cases may be considered. Where the half-life of the daughter is slightly shorter than that of the parent, the daughter's activity starts from zero, rises to a maximum, and then seems to decay with the same half-life as that of the parent (figure A3.2). At this point the daughter is disintegrating at the same rate as it is being produced and the two isotopes are said to be in a state of *transient equilibrium*. In quantitative terms,

$$A_B = A_A \frac{\lambda_B}{\lambda_B - \lambda_A} \qquad (A3.3)$$

239

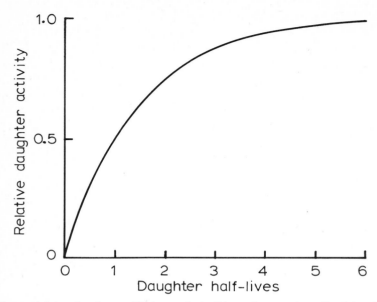

Figure A3.1.   Secular equilibrium – the build up of a very short-lived daughter radionuclide from a much longer lived parent radionuclide.

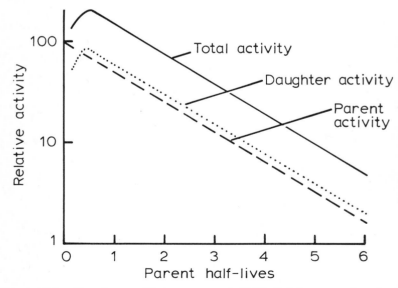

Figure A3.2.   Transient equilibrium –parent half-life slightly greater than that of the daughter.

The total activity arising from radionuclides which tend to a state of transient equilibrium reaches a maximum value which differs in time from that of the daughter. The time when the total activity is at a maximum (t_{mt}) may be derived by

$$t_{mt} = \frac{1}{\lambda_B - \lambda_A} \log_e \left[\frac{\lambda_B^2}{2\lambda_A\lambda_B - \lambda_A^2} \right] \qquad (A3.4)$$

Secondly, in the case where the half-life of the daughter exceeds that of the parent, no equilibrium between parent and daughter is possible

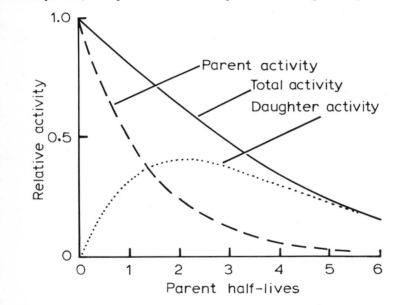

Figure A3.3.   No equilibrium – daughter half-life is greater than that of the parent.

(figure A3.3). The daughter activity reaches a maximum at a time (t_{md}) which may be calculated from

$$t_{md} = \log_e \frac{\lambda_B/\lambda_A}{\lambda_B - \lambda_A} \qquad (A3.5)$$

Eventually a point is therefore reached where the daughter decays at its own characteristic rate; the parent, because of its shorter half-life, having decayed away. The total activity in this case does not increase to a maximum but decreases continuously.

Appendix 4. Units and definitions used in radiological protection

Units

Becquerel	(Bq) – unit of radioactivity – equal to one nuclear transformation per second. (1 Bq = 2·7 × 10^{-11}Ci).
Gray	(Gy) – unit of absorbed dose – equal to 1 J kg^{-1}. (1 Gy = 100 rad.)
Sievert	(Sv) – unit of dose equivalent. (1 Sv = 100 rem.)
Roentgen	(R) – unit of exposure – equivalent to 2·58 × 10^{-4} C kg^{-1}.

Definitions

Absorbed dose (D) – quantity used to measure the energy deposited per unit mass

$$D = \frac{\mathrm{d}\bar{\varepsilon}}{\mathrm{d}m}$$

where d$\bar{\varepsilon}$ is the mean energy imparted by ionizing radiation to the matter in a volume element, and dm is the mass of the matter in that volume element. The unit of absorbed dose is the gray.

Relative biological effect (RBE) – the ratio of the amount of energy of any radiation required to produce a given effect to the energy required of 200 keV X-rays to produce the same effect.

Quality factor (Q) – a factor which is intended to allow for the effect on the detriment of the microscopic distribution of absorbed energy. It is numerically defined as a function of the collision stopping power in water at a point of interest. Effective quality factors are given in table 1.2.

Dose equivalent (H) – the dose equivalent, at a point in a tissue, is defined by

$$H = DQN$$

where D is absorbed dose, Q is the quality factor, and N is the product of all other modifying factors – for example absorbed dose-rate and fractionation. The unit of dose equivalent is the sievert.

Weighting factors (W_T) – a factor representing the proportion of the stochastic risk resulting from irradiation of tissue T to the total risk when the whole body is irradiated uniformly. Weighting factors for different tissues are given in table 3.8.

Effective dose equivalent (H_E) – a sum derived by

$$H_E = \Sigma_T \, W_T \, H_T$$

where W_T is a weighting factor, defined above, and H_T is the dose equivalent in tissue T. The summation is carried out over the same period for all tissues. The effective dose equivalent acts as an indicator of the risk of death from somatic effects, and of hereditary effects in the first two generations, which are assumed to result from any radiation, whether uniform or non-uniform, from both external and internal sources. It is not a complete indicator of health detriment, however, because it does not allow for non-fatal somatic effects, and does not include hereditary effects in subsequent generations.

Committed dose equivalent (H_{50}) – the integrated dose equivalent in a particular tissue which will be received by an individual following an intake of radioactive material into the body, the integration time being set at 50 years – to correspond to a working life-time. It is defined by

$$H_{50} = \int_{t_o}^{t_o + 50\,y} \dot{H}(t)\,dt$$

for a single intake of radioactivity at time t_o, where $\dot{H}(t)$ is the dose equivalent rate in the organ or tissue.

Committed effective dose equivalent – a quantity representing the total risk of specified somatic and hereditary effects to an individual and his progeny, as a result of an intake of radioactive material, including the risk from irradiation in subsequent years resulting from the intake. It is derived by multiplying the committed dose equivalents (H_{50})s to the individual tissues, as a result of intake, by the appropriate weighting factors (W_T), and summing the results.

Annual limits of intake (ALI) – secondary limits which state the maximum yearly intake for individual radionuclides. They will normally correspond to a committed dose equivalent, from an intake of a given radionuclide, that is equal to the appropriate dose equivalent limit set for workers, or for members of the public. If an 'average' individual were to take in a given radionuclide at the appropriate ALI each year, for 50 years, his dose equivalent during the 50th year would be equal to the appropriate dose equivalent limit.

Annual dose equivalent – the dose equivalent received during a year. For external radiation this is the sum of the dose equivalents received during the year, but for the internal irradiation of a particular organ or tissue it is the integral of the dose equivalent rate in that organ or tissue during that year. The dose equivalent rate may therefore also include a component due to radioactivity remaining within the body as a result of intake during previous years.

Collective dose equivalent (*S*) – a quantity which relates the total exposure of a group of individuals to a particular source of radiation exposure. It can be defined in two ways: either as

$$S = \Sigma_i \bar{H}_i N (\bar{H})_i$$

where $N(\bar{H})_i$ is the number of individuals in population subgroup i, receiving an average dose equivalent of $\bar{H}_i$; or

$$S = \int_0^\infty H N (H) \, \mathrm{d}H$$

where $N(H)\mathrm{d}H$ is the number of individuals receiving a dose equivalent between H and $H + \mathrm{d}H$. The collective dose equivalent is a quantity which can be used in cost–benefit analyses for the purpose of justification and optimization; its unit is the man-Sv. The effective dose equivalent may be substituted for the dose equivalent to give a **collective effective dose equivalent**.

Per caput **dose equivalent** – an 'average' dose equivalent. It can be obtained either by dividing the collective dose equivalent in a population by the number of individuals in it, or by calculating the average exposure to, or intake of, the radioactivity released from a source, and thus the average dose equivalent. The effective dose equivalent may be substituted for the dose equivalent to give a *per caput* **effective dose equivalent**.

Dose equivalent commitment (H^c) – the integration of the *per caput* dose equivalent rate, $\bar{H}(t)$, from a given source as a function of time.

$$H^c = \int_0^\infty \bar{H} (t) \, \mathrm{d}t$$

It presupposes that it is possible to calculate the variation of *per caput* dose, from a given source of radiation, as a function of time. If the integration is terminated at a specified time, instead of infinity, then the quantity is termed a *truncated* or *incomplete* dose equivalent commitment. The H^c can be used to estimate the maximum future annual *per caput* dose equivalent from a continued practice, which can then be compared with other values. The effective dose equivalent may be substituted for the dose equivalent to give an **effective dose equivalent commitment**.

Collective dose equivalent commitment (S^c) – the integration of the collective dose equivalent rate, $\dot{S}(t)$, from a given source as a function of time

$$S^c = \int_0^\infty \dot{S}(t)\, dt$$

This quantity, therefore, expresses the total collective dose equivalent from a given practice for use in justification and optimization studies. As with the dose equivalent commitment the integration can be terminated at a given time, to give a *truncated* or *incomplete* quantity. The effective dose equivalent may be substitued for the dose equivalent to give a **collective effective dose equivalent commitment**.

Recommended Reading

Glasstone, S., 1967, *Sourcebook on Atomic Energy*, 3rd edn. (New York: Van Nostrand.)

Bacq Z.M., and Alexander, P., 1967, *Fundamentals of Radiobiology*, 2nd edn. (Oxford: Pergamon Press.)

Cember, H., 1969, *Introduction to Health Physics*. (Oxford: Pergamon Press.)

Spiers, F.W., 1968, *Radioisotopes in the Human Body*. (London, New York: Academic Press.)

Coggle, J.E., 1971, *Biological Effects of Radiation*, Wykeham Science Series. (London: Taylor and Francis.)

Recommendations of the International Commission on Radiological Protection No. 22 Implications of Commission Recommendations that Doses be kept as Low as Readily Achievable; 1973. (Oxford: Pergamon Press).

No. 26 Recommendations of the International Commission on Radiological Protection; 1977. (Oxford; Pergamon Press)

No. 27 Problems Involved in Developing an Index of Harm; 1977. (Oxford: Pergamon Press)

Sources and Effects of Ionizing Radiation, 1977, United Nations Scientific Committee on the Effects of Atomic Radiation, Report to the General Assembly; United Nations Publication E.77. IX.1.

Royal Commission on Environmental Pollution, 1976, Sixth Report: Nuclear Power and the Environment, Chairman, Sir Brian Flowers (London: HMSO).

Eisenbud, M., 1973, *Environmental Radioactivity*, 2nd edn. (London, New York: Academic Press).

Eichholz, G. G., 1976, *Environmental Aspects of Nuclear Power*, (Ann Arbor Science Publishers).

Radioactivity in the Marine Environment, 1971, Chairman A. H. Seymour; (US National Academy of Sciences).

Effects of Ionizing Radiation on Aquatic Organisms and Ecosystems, 1976, Technical Reports Series No. 172; (Vienna: International Atomic Energy Agency).

Index

THE WYKEHAM SCIENCE SERIES

(Volumes 5, 18, 20 and 24 out of print)

† (*Paper and Cloth Editions available.*)

THE WYKEHAM ENGINEERING AND TECHNOLOGY SERIES

SOCIAL SCIENCE LIBRARY

Manor Road Building
Manor Road
Oxford OX1 3UQ
Tel: (2)71093 (enquiries and renewals)
http://www.ssl.ox.ac.uk

This is a NORMAL LOAN item.

We will email you a reminder before this item is due.

Please see http://www.ssl.ox.ac.uk/lending.html
for details on:

- loan policies; these are also displayed on the notice boards and in our library guide.

- how to check when your books are due back.

- how to renew your books, including information on the maximum number of renewals.
  Items may be renewed if not reserved by another reader. Items must be renewed before the library closes on the due date.

- level of fines; fines are charged on overdue books.

Please note that this item may be recalled during Term.